The Inner World of Research

The Inner World of Research

On Academic Labor

Stefan Svallfors

A

ANTHEM PRESS

Anthem Press
An imprint of Wimbledon Publishing Company
www.anthempress.com

This edition first published in UK and USA 2021
by ANTHEM PRESS
75–76 Blackfriars Road, London SE1 8HA, UK
or PO Box 9779, London SW19 7ZG, UK
and
244 Madison Ave #116, New York, NY 10016, USA

First published in the UK and USA by Anthem Press in 2020

British Library Cataloguing-in-Publication Data
A catalogue record for this book is available from the British Library.

Library of Congress Control Number: 2021937604

ISBN-13: 978-1-83998-157-9 (Pbk)
ISBN-10: 1-83998-157-1 (Pbk)

This title is also available as an e-book.

He was never afraid when writing, but only then.
− Per Olov Enquist, *Liknelseboken*
[The Book of Parables] (Stockholm: Norstedts, 2013)

I have seen things you people wouldn't believe. Attack ships on fire off the shoulder of Orion. I watched C-beams glitter in the dark near the Tannhauser Gate. All those moments will be lost in time, like tears in rain.
− The replicant Roy Batty's monologue from the film *Blade Runner* (1982)

To fall is to understand the universe.
− Sara Stridsberg, *Beckomberga: ode till min familj* (Stockholm, Albert Bonniers förlag, 2014), p. 354. Extract from *The Gravity of Love* translated by Deborah Bragan-Turner

CONTENTS

ACKNOWLEDGEMENTS

Writing this book has been a solitary exercise, based largely as it is on my own experiences and memories. Nevertheless, there are many people who have contributed to it, nay, made it possible.

Gunilla Gerland, Olle Häggström, Mattias Marklund, Lars Nyberg, Bo Rothstein, Chris Reus-Smit and Joel Sundin accepted to be interviewed for the book and provided many insights into their worlds of thoughts and experiences. The same goes for the interviewees who preferred to remain anonymous.

Joa Bergold, Erica Falkenström, Janne Flyghed, Gunilla Gerland, Anne Grönlund, Olle Häggström, Tomas Lappalainen, Christer Nordlund, Bo Rothstein, Hugo Svallfors and Signe Svallfors provided essential feedback on previous Swedish versions of the text. Their comments stretched from the positively enthusiastic to the courageously destructive. The result is a much better book than I could ever have achieved myself.

Jens Beckert and Monika Kostera encouraged me to publish the text in English and have been helpful in navigating publishing houses.

Neil Betteridge translated the text with flavour and elegance. It was truly miraculous to see my text transformed into a different language without any effort on my part.

Erica, for making everything possible. She is a singular person who has chosen to share her singularity with me.

THE CATHEDRAL ON THE PLAIN

An old legend tells of how Chartres cathedral was struck by lightning and burned to the ground. People flocked in their thousands from different directions like a vast procession of lemmings from all four corners of the world. All kinds of people came and together they rebuilt the cathedral on the old foundations. They lived their lives by the enormous edifice until it was complete – master builders, labourers, artists, jesters, nobles, prelates and citizens – but they remain anonymous; to this day, no one knows the names of those who built the cathedral of Chartres.

[…] So if someone asks me what I would like the purpose of my films to be I could reply: I want to be one of the artists in the cathedral on the great plain. I want to sculpt from the stone a dragon's head, an angel or a devil, or perhaps a saint, it matters not which; I derive such satisfaction from all manner of things. No matter if I'm a believer or not, no matter if I'm a Christian or a heathen, I help to build the cathedral because I've learnt to form faces, limbs and bodies from stone.[1]

I also want to help build the cathedral on the great plain. I want to use my abilities because I have them, because I am good at what I do and because it gives me a deep sense of satisfaction. I never think of 'citation index' or 'impact factor' when I solve research problems or translate nebulous thoughts into coherent text. What I think of is how the dragon's head can emerge from the stone.

These qualities – my professional aptitude and professional pride – I want to safeguard. Cultivate, exercise, talk about. I want to shield them from the malice that can blight academia, malice by which people impede and oppress each other. Destroy each other's potential. I also want to shield them from those who seek to turn us into pawns in the knowledge war, fought in order that our research nation may triumph (over whom and why is unknown).

It is not my intention to make my own lifestyle somehow prescriptive. Maybe you, dear reader, wish to live a completely different life to the one I have done and found blessed. But nor do I want my professional experiences – of joy and suffering, of euphoria and despair – to melt away like tears in the rain.

Hence this book.

This book has its origins in some of my burgeoning frustrations. The first concerns how we university scholars do not teach about the realities of research. Perhaps it is unteachable, but we should at least talk about what we do and how it feels. Such conversations are not had – other than in passing with individual colleagues. Nor are we capable of telling those who finance and try to control our research what we do and what we need to do it well.

The second is about the relationship between body, emotion and knowledge. The intellectual life is regarded alarmingly often as a non-physical, emotionless activity – as if our thoughts and our knowledge were disengaged from our bodies and their needs, desires and limitations. I believe that this view gives rise to a narrow, too regulated, emotionally amputated way of approaching the objects of our knowledge. And that this inhibits our creativity as researchers and as human beings.

My third frustration is about how the inter-human relations in which knowledge production is embedded are so rarely given the recognition they deserve. The solitary researcher – the genius in his chamber – is often pitted against vast collaborative environments in which hundreds of researchers work on gigantic projects. But it is neither the lone scientist nor the huge centres that are the most relevant level of knowledge generation. This particular honour is reserved for the group, the team, and its cognitive, emotional and social relations. A mini-collective, a micro-cosmos that is virtually invisible when its research is presented, debated and financed.

My final and perhaps greatest frustration concerns just this: research financing and research policy. One often forms the impression that politicians and civil servants think that good research can be administered by decree. As if the researchers were pliant soldiers in a knowledge war. Assembling creative and productive research environments is in fact a delicate and emotional enterprise. It is about building (literally and figuratively) spaces in which people feel secure and remember all they know – where they have the courage to share their agonies and their joys.

To tackle these frustrations, I am adopting an at once modest and daring strategy, one that is expressed, for example, in the way I rarely make reference to any literature apart from when recounting ideas and arguments that I have borrowed directly from someone else (and remember that I have done so!). I have not read any other essay like this one, and I have made no systematic review of the writings published on knowledge production and its organisation. To be sure, I have read a great many texts related to my intellectual quandaries, but the selection has been unmethodical, one could say almost coincidental and random. The texts might often have been about very different subjects than researchers and research, but they have all addressed the conditions and limitations of creativity in a way that has reduced actual disciplines to relative insignificance.

So the book can hardly be classified as research; it is difficult to even pinpoint a genre to which it might belong. I think of it – half in earnest, half in jest – as an 'auto-anthropological voyage of discovery'. By this I imagine that in using my own scientific achievements and biography as my point of departure, I can reflect things which I think others might recognise. Or failing that, be forced to describe their own everyday life as scientists and discoverers in their own words. This book is not just about me, but there is no one it is more about than me. The northerner in me winces at this embarrassing self-absorption, but I have learnt to ignore this Jante-law-driven surliness.

The book is therefore based very much on my first-hand experiences in the frontiers, fringes and borderlands of knowledge production, on my reflections about the joys and discomforts of being a researcher. On memories and experiences accumulated over three decades of practical research, during which, for both good and ill, I have come to better know myself and the environments in which I work. I will talk about the demons I have met and how I have fared in my battles with them. I recall moments of euphoria and despair. We could do with more such self-revelatory observations – scientists like to present an image of themselves and their work as being a little better and a little more polished than they actually are.

But this is a risky approach. When I repeatedly find that I cannot remember in detail the books I read just a few years ago, I am reminded – correctly, given how the memory adds and subtracts in the most unreliable way – that childhood memories, naturally, can also not really be trusted. This is how I remember things, but how much has been added and subtracted by the active memory and how much reflects what actually happened and what I felt about it is impossible to say.

When it has all felt a little too pretentious, I have looked at myself in the mirror and asked myself about my self-centred narrative: *if not you, then who?* In his autobiography, *Vänsterdocenten*, philosopher Torbjörn Tännsjö has Stendhal reassure us that 'an impassioned individual is one who is so obsessed with something that he is prepared not just to die or kill for it, but to make a fool of himself for it'.[2] I am not prepared to die or kill for what I am attempting to set forth in this book, but I am prepared to make a fool of myself.

The book is also based, however, on conversations and interviews I have had with friends and colleagues who have had astute things to say – in various forms, from fairly structured interviews to informal chats on the book's themes. While the overriding topic of these conversations has been the organisation of research, they have touched upon everything from the deep structures of the brain to the blunders of research policy. Sometimes, these conversations have confirmed what I already thought or suspected myself; sometimes they have made me see things in a fresh light or given me cause to re-evaluate my own ideas.

Through these conversations I have gained access to experiences and insights that I myself have lacked and seen my own moments of elation and shortcomings reflected in the similar experiences of others. Theoretical physics, neuroscience, mathematical statistics, baroque music, painting and empirical social science is an incomplete list of the fields and perspectives into which these interviews have delved.

I believe that what I have found in these conversations is a kind of fundamental grammar of research, or perhaps even the social and emotional dimensions of intellectual life. What I have tried to do in this book is to put into words some aspects of this grammar – as a guide, a diversion and maybe a comfort for my fellow travellers on this destination-less journey upon which we have embarked.

Everything's connected is perhaps the most important message, if at all there is any. The way we think, write and speak is intimately bound to our own bodies and their comforts and discomforts. How we construct our research objects and analyse them is the product of deeply embodied pattern-recognising knowledge. This embodied knowledge is, in turn, closely tied to the physical and social environments that enclose us, and the emotional and cognitive relations that are cultivated and nurtured there. What we dare and dare not do, assuming we can allow ourselves full access to our resources, is determined by the group's social climate. And the group's constitution and dynamics are moulded by research leaders and research financing, which can facilitate or frustrate the micro-processes that decide the quality and relevance of the knowledge generated.

What matters is just this: having access to all our resources – at the moment of lecture, at the time of writing, in discussion with the others – and helping each other to recover these resources, this inner galaxy of linked abilities. What people can achieve together when working at the peak of their creative abilities is altogether miraculous; there is no other word for it. Research leadership is about making it easy for us to perform these miracles, whether it be in the team's local workaday environment or under the broader conditions of research policy.

What is at stake? Human creativity. What we need to do is make optimal use of human creativity, to not let time and energy be wasted on trivialities and fruitless frictions. To be what creativity researcher Ken Robinson might call 'in the Zone' most of our time.[3] In the Zone, where we are one with our element, where time seems to stand still, where we can exploit all our faculties to the full. Where there is 'perfect peace mingled with an excitement that strains every nerve to the utmost', as flight pioneer Wilbur Wright succinctly put it.

For this is the place where we constantly yearn to be. When we are gripped by our own exposition during a lecture, reach our audience and carry an entire room. When in conversation with a colleague we engage in give and take at precisely that level of insight and proficiency we are able to reach. When our collaboration produces results that none of us could have achieved alone. When the text appears, as if by magic, on the screen right in front of our astonished eyes. These moments of unsullied happiness, at the verges of our own abilities. When we remember all we know.

There are many ways of misreading this book. One is to think that my criticism of overly regulated and indifferent research implies a belief that research and its conclusions materialise as if through divine inspiration or mystical intuition. This is not at all the case. There are no shortcuts to intuitive, embodied, deeply familiarised knowledge or to a realm of ideas that integrates emotion with reason. Before one can break the rules constructively, one must first know them intimately. No pain no gain, as the old saying goes, a maxim to which all creative activity submits.

Neither shall my claims about the embodied and the emotional be inferred as some kind of relativism. I am an epistemological realist. I hold that there are phenomena, forces and mechanisms that lie beyond the observer or the investigator. They were there before they were observed by the researcher. They thus have an 'objective' existence in the sense that they are not called into being by the very act of observation. What I try to capture in this book is, instead, about how one attains knowledge of this from a reality that exists, in part, independently of scientific practice. And the role that physiology, emotion and sociality play in this.

Another misinterpretation would be that I want academia to be isolated from the rest of society. That I just want to be left in peace. That is also incorrect. The fields of inquiry to which social scientists devote their research are inevitably political. In the sense both that they arise in a political and conflicted context, and that their results have ramifications on the society studied.

All I want is for us to be governed by a deeper understanding of how knowledge is actually produced and of how the people who produce it function and what propels them. I very much doubt that this will be the case. The opposing forces are so strong. But one can always hope.

Chapter 1

ALONE TOGETHER

All of old. Nothing else ever. Ever tried. Ever failed. No matter. Try again. Fail again. Fail better.

First the body. No. First the place. No. First both. Now either. Now the other. Sick of the either try the other. Sick of it back sick of the either. So on. Somehow on. Till sick of both. Throw up and go. Where neither. Till sick of there. Throw up and back. The body again. Where none. The place again. Where none. Try again. Fail again. Better again. Or better worse. Fail worse again. Still worse again. Till sick for good. Throw up for good. Go for good. Where neither for good. Good and all.[1]

The Body as Knowledge Receptacle

Half an hour to go before my turn. I'm not well. Can feel the nausea coming in waves. Why did I accept? Why subject myself to this? What am I actually doing in Seville? And what kind of insane idiots fly me across Europe to hold a talk? Don't they see that I have nothing to offer?

I thumb through my lecture notes. So flat. Boring. Impossible to get anyone engaged in. I'm not even interested myself. And so tired. My mouth feels full of gravel, or cotton. My Dutch colleague Harry comes walking towards me. He smiles cheerfully when he catches sight of me. We exchange a few words. To my surprise I sound happy, calm, modest and self-assured. I feel dizzy. As if I'm standing beside myself talking.

Up on the podium my body is still in protest mode. But it does actually start to relax. I run my gaze over the audience and the hall we are in. Search out some pleasant looking faces to fix my eyes on when I talk. Should one of them look uninterested when I start, I'll just move on to another. Someone who's keeping up, nodding encouragingly. Someone to steal energy from.

It's starting. I'm introduced in Spanish before we switch to English. I start to talk. I've got their attention. I'm calm, my body feels warm. I find a woman down to the left who keeps nodding; far to the right, the guy from Barcelona is looking genuinely interested. In the middle an alert-looking student. I avoid Harry; he always looks so sceptical.

Done. A bit slow in the middle, otherwise flowing. Then the Q&A session. Good questions. Engaged, sometimes hard, but interestingly so. The guy from Barcelona has an especially good comment. I head off in long, complex but completely coherent answers. I don't need to defend myself. I know it worked, that they're all on board.

So it's over. Some of them approach the podium, want to carry on the discussion, exchange cards. Head all bubbly now. Feel happy. Almost euphoric. I should do this more often. It's so much fun! Being in the centre, elevated but still one with the others. I remember a little vaguely that it felt terrible just a short while ago but can't for the life of me conjure up the feeling again. This is living life to the full.

We have bodies. The things we think, feel and know reside in the body. That stands to reason, surely. Where else would they be?

It is just that researchers often think about thinking and the intellect as if we lacked bodies. As if thoughts were in an ethereal space or on the disembodied brain we encounter in the P. C. Jersild novel *A Living Soul* or Dennis Potter's TV drama *Cold Lazarus*. And that the intellectual occupations are organised as if we had no bodies, as if how we think is not influenced by how our bodies feel. Creativity researcher Ken Robinson talks sarcastically about how professors seem to regard their bodies 'as a form of transport for their heads [...] as a way of getting their heads to meetings',[2] when our bodies are actually inseparably amalgamated with our thoughts.

I'm reminded of this every time I'm due to give a lecture. About how the purely physical conditions of the lecture bear upon my physical well-being and how, in turn, this affects my intellectual performance.

I like standing when I have to speak. I want a proper pulpit to declaim from, big enough to take all my papers and to spare me the hassle of going back and forth to change slides or whatnot. This makes me lose my thread. The room should be rather wide and not so deep. And I should be close enough to the front row for the atmosphere to be 'tight'.

Ideally, the room should be full. I dislike talking to a room that is half-empty and where the audience is sitting far back. The barriers grow and it's hard to connect. Not to mention the awkwardness that descends when equipment fails to work, when screens freeze and the projector goes on strike. At times like that it can be almost impossible to pull things off. You feel sick and your brain stops working.

Conversely, when things are flowing: 'The better it goes, the more physical I become', as one of my lecturing friends put it. 'I perform with my hands, in the end I'm so close to the front row that I'm almost touching the audience.'

However, even if it is in my role as lecturer that I am most acutely reminded of the body's role in the intellect, exactly the same thing applies to when I am reading or writing. The (dis)comfort and rhythm of the body in the endeavour are inseparable. Writing is a deeply embodied phenomenon, 'a hugely physical experience', as one of my writer friends expresses it. 'I can't write in my office,' she continues. 'I must sit in my armchair, a really unergonomic one, with my papers spread out in front of me, otherwise nothing happens. I need time to hear my own voice, where words beget words, where everything flows, where only later can I see what came of them.'

For others I have spoken with, the purely physical circumstances of their writing is of less significance. When thinking, political scientist Bo Rothstein tries to find a 'line of argumentation' with which to structure a text, which subsequently arrives in a mysterious flow that pays little heed to place and position. How this actually works is something he does not care to reflect much upon: 'I'm afraid of over-analysing it, for if I start to think too much about how I go about writing, it won't work anymore. It's like riding a bike; you just do it, but if you start thinking too much about the fact that you're cycling, you can lose your balance.'

Physicist Mattias Marklund says that the first versions of his texts usually come about with pen and paper. 'Firstly, it feels as if it's a physical object that I won't misplace,' he smiles. 'Secondly, there's also a pace to it that suits the thought process. And then when you're sitting at a computer, there can be a certain technical frustration that hampers your writing. Paper and pen, that's *foolproof*.'

Thus equipped he sets to work translating the 'algebraic expressions' he has in his head, more or less formulaic images of what he wants to say, into words. Whoever makes high-tech associations on hearing the terms 'plasma physics' or 'quantum vacuum' should picture Mattias Marklund with just a pen and paper as his tools of thought.

In their *Mind over Machine: The Power of Human Intuition and Expertise in the Era of the Computer*, brothers Hubert (the philosopher) and Stuart (the computer scientist) Dreyfus discuss human learning and how it differs fundamentally from how computers operate.[3] They identify five different stages of mastery, from complete novice to expert. In the first three stages, the beginner is transformed into a competent rule-follower. At the first stage, a beginner learns a few simple rules of thumb; at the second, they learn to apply them to different situations. When they are at the third stage, they are not only able to apply rules in different situations but can also see the implications these applications have for the outcome, and adjust their rule-following accordingly. These three levels are, then, based on rule-following and conscious decision-making. Such knowledge can be codified and fed into a computer programme or gathered in expert systems. Or in textbooks on social science methodology.

But then something happens. At the top two stages of mastery, when the competent rule-follower becomes first 'proficient' then 'expert', what they know can no longer be codified. Here, knowledge is based on deep familiarity, a feeling or intuition that's all but physical and difficult to account for. They just 'know what to do', they just 'feel at home' in situations. The virtuosity that people can demonstrate when in full mastery of something is an immediate embodied experience, a way of being in the world rather than a way of thinking.

No matter if the virtuosity concerns football or research, the processes are the same. When the brilliant football player suddenly sprints off in an unexpected direction, sees or rather senses a space where no one else does, it is principally the same as when the researcher suddenly notices a pattern that indicates a new discovery. It is knowledge that stems from having been in similar situations so many times before that the opening becomes immediately visible, the nature of the discovery obvious. It is a form of knowledge not based on rules and conscious decision-making. Not even afterwards is it possible to pinpoint what gave rise to the action or discovery. It was just there.

Research done in recent decades on the cerebral neural system adds some interesting aspects. Here, neuroscientists have demonstrated the presence of two distinct systems in the brain. The 'hot' system is intuitive, emotion-driven and fast. It is rooted in the biologically older part of the brain and transmits an instantaneous, emotion-based signal when faced with certain events and situations.

The other system is the 'cold' system, and it is analytical, cognitive and relatively slow. It is rooted in the neocortex, which is the most recently evolved of our brain structures and the home of our reflective intellect. It is here where we take conscious decisions.

They are also referred to by neuroscientists as the brain's 'explicit' and 'implicit' systems.[4] The former is controlled, cognitive, logical and emotionally detached; it is also relatively slow and effortful. The latter is intuitive, associative, emotional and embodied. It is fast too: one just knows, immediately, where the ball is heading, what part of the table is interesting.

So it seems as if the two highest stages of the Dreyfus developmental model relate to the gradual embedding, in some people and in some respects, of what are initially conscious processes of the brain's cold system in the hot, implicit and rapid intuitive system. So one 'does the right thing' only because one 'knows what to do'.

In a *New York Review of Books* article, chess virtuoso Garry Kasparov describes the battle between human and machine as fought out on the chess board.[5]

These days, the most powerful computer programs can beat the best human players. But it took an astoundingly long time for this achievement to come about. And it is still through pure number crunching that the computers win, not by 'thinking' more intelligently. But they do not win by calculating precisely all possible moves at every turn; this would exceed the capabilities of even the most powerful computer. As long as the programmes worked in this way, the grandmasters could keep them at bay. Instead, the best programmes now have built-in shortcuts, shortcuts designed by humans, humans combining deep familiarity-knowledge of chess and of coding.

The story of how the chess computers finally defeated a human being is thus testimony, Kasparov seems to be arguing, not to how machine finally masters man, but to how man can have mastery of his own intellectual power. Beyond calculation, we find human creativity.

In *Blink: The Power of Intuitive Thinking* Malcolm Gladwell tells the fascinating story of when the Getty Museum sought to purchase a Classical Greek statue.[6] Everything seemed in order. Chemical analyses showed that the material of the statue and the soil fragments trapped in its crevices tallied with its alleged provenance. The documentation of its origins was immaculate. Yet the museum wanted to check a little more closely; after all, it was just about to part with tens of millions of dollars.

So it called in some experts to look at the statue. And they immediately felt that something was wrong. *Felt* is the right word, for what they described was an almost physical discomfort at the sight of it. Even though everything seemed kosher, even though all the papers and analyses were in place, they had a gut feeling that something was wrong. The statue felt 'fresh', as one of them put it. Another got snagged by the fingernails. They just felt wrong.

Sure enough, the statue and the record of its discovery proved to be a cunning forgery. But what is of interest to us is the immediate, almost intuitive, thought-feeling that made the experts recoil before the statue, the manifestation of a virtuosity strongly reminiscent of the highest stages in the Dreyfus model. Where one 'just sees', 'just knows', beyond the controlled, indeed, almost beyond thought and word.

Gladwell calls this ability *thin-slicing*: to instantly see, with thin and scant information, what aspect of a situation is relevant. The cognition research that he bases his ideas on here – conducted by Gerd Gigerenzer and colleagues at the Max Planck Institute in Berlin – talks instead of 'heuristics', simple rules of thumb that are often much more fruitful than elaborate decision-models.[7]

This implicit knowledge is profoundly contextual. It is not based on general prototypes and rules; it is based instead on implicit memories of concrete situations and unique examples.[8] Consequently it takes a long time to amass, to give us time to explore a domain so that we may assimilate its embodied knowledge. Once there, it is incredibly robust, unlike explicit rules and details that are easily forgotten or distorted; for implicit knowledge is a kind of pattern recognition that derives from having experienced similar situations time and time and time again. And it is critical to our ability to focus on the right things in a situation and see immediately what is relevant.

What *Blink* therefore does not really explain is that this way of understanding does not come free, but is the result of thousands of hours of drudgery. It comes from having been in similar situations so many times before, of having seen similar things time and time and time again. So that the knowledge eventually becomes established in the brain's emotional and rapid system, which lets us instantly see and understand things. This process very much engages our physical and emotional selves. 'It has to hurt,' as the Dreyfus brothers put it when describing the conversion of controlled knowledge into a deeply embodied ability.[9]

So there are no shortcuts, no way of skipping the drudgery to pass directly to virtuosity. It was only because they had seen thousands of other statues and statue fragments, because they had read up on the subject and read up again, because of all the hours they had stared and read and thought, that the Getty museum's co-opted experts could instantly feel that something was wrong. For me, the statue would have just been a piece of marble in the shape of a person.

Poet and author Bob Hansson tells the story of when a rock journalist interviewed guitarist Jeff Beck and asked him the ingratiating question, 'How do you get to be as good a player as you?' Beck replied, 'You get yourself a guitar when you're like twelve and then play every day for like forty years.' 'No shortcuts,' asks the horrified journalist. 'No shortcuts.'[10]

No shortcuts. In the sequel *Outliers: The Story of Success*, Gladwell writes how the route to virtuosity takes 'ten thousand hours'.[11] What distinguishes the virtuoso from the merely competent is above all the simple fact that they have dedicated themselves more to what they are good at. It matters not if you are a hockey player, a chess whiz, a sociologist or a guitarist. Ten thousand hours. Then you can sense what is right and wrong in that intuitive way that refuses to be codified in rules.

An important difference between explicit rule-following and implicit embodied intuition lies in the way they can be disrupted. In one fascinating study, a group of very accomplished amateur golfers were asked to play first in peace and quiet and then against loud background music and noise.[12] Their game deteriorated markedly with the noise. The same experiment was then repeated with a group of professional golfers. Their game actually *improved* by being disturbed! The explanation is that the amateurs were still at the rule-following stage in which they needed their conscious mind to perform.

When disturbed, they lost concentration and their game suffered. As for the professionals, however, their playing had shifted into the brain's intuitive system and it was advantageous to have the brain's conscious system otherwise occupied by the music, thus mitigating the danger that it would disturb the brain's intuitive, emotion-based system.

In another fascinating experiment, the Dreyfus brothers had a chess grand-master play lightning chess against a skilled opponent *at the same time* as he had to do complicated mental arithmetic.[13] His powers of calculation were thus fully taken up by something other than chess. Despite this, he won easily; for a chess player at his level, the game has transferred to the implicit system, where it is no longer disturbed by mental operations in the explicit system.

As Swedish educational researcher Lars-Erik Björkman points out, the implicit system can, however, be easily disrupted by emotional and physical dis-comfort.[14] Under stress and in demanding situations, we often shift to rational and explicit memories and solutions. Like when I suddenly stop trusting my abilities during a lecture and start to read aloud from my notes. It becomes mechanical, lifeless. If I had just had confidence in my own improvisations, I would have fared much better. And here is where the spatial framing of intel-lectual meetings plays a crucial part. In uncomfortable rooms that are physic-ally disagreeable to be in, meetings become controlled, frigid, anxiety-ridden. And thoughts muddier, arguments poorer.

In the middle of writing, I am physically and brutally reminded of how closely body and mind are integrated. I have an accident that leaves me with a complicated elbow fracture. For several weeks I can't write, and after that only with my right, uninjured hand (which happens to be my favoured one anyway). Or more accurately: I can't write, at least not the kind of text that requires some mental effort. I can't even think coherently. As if my brain was in my elbow. That mystical flow that powered the genesis of the text had been stemmed by a broken elbow. If it were the case that one composed the text in one's head to then commit it to print, the absence of a hand would not have been such a handicap. It would have just retarded the process a little. But the text and the thought engage in a mystical dance involving brain, body and computer. And a broken elbow suddenly makes it impossible to string words into meaningful paragraphs.

Stuart Dreyfus, who apart from being a prominent computer scientist is also a pretty decent chess player, is asked in an interview with Danish political scientist Bernt Flyvbjerg to identify where in the body a chess player feels a move is right.[15] In the head or some other place? 'In the whole body,' replies Dreyfus. 'In the pit of my stomach.' But this is like asking where you feel hungry when you are hungry, he continues: you experience your whole body as hungry and it is exactly the same for the chess player. The deeply rooted knowledge creates physical sensations of a real and almost disagreeable kind.

Dreyfus himself has had to accept that as a chess player he will never be more than a skilled rule-following amateur and that he lacks the ability to level up to a grandmaster, where one feels and lives the game. But in a way, he admits, this was lucky, because it was this that taught him to see the limitations of the analytical, rule-following way of thinking. And to understand that there were other, higher forms of skill.

That thinking is deeply embodied is also something observed by medical researchers, or rather researchers who look into how doctors experience situations and take decisions. Taking central place here is the term 'gut feeling'.[16] Almost all doctors faced with making a decision – sometimes a life-and-death one – report that they often let themselves be guided by their 'gut feeling', a hunch for what is right in a given situation that eludes definition and that is based on the kind of deep knowledge that I have been trying to pin down here.

Therefore, these doctors, as well as those who have studied the significance of gut feelings, are highly sceptical about explicit decision-making rules and the growing influence of cost-benefit calculations in the healthcare sector. Their judgement and ethical responsibilities, they argue, risk being disregarded, and this to the detriment of their profession. 'I was just following the rules', a doctor can say when misfortune strikes, even though she had felt, physically, that what she was doing was wrong.

Here, however, we need to distinguish between a gut feeling that is genuinely based on deep-seated familiarity-knowledge and one that is just engendered by prejudice, fear and antipathy. In her thesis on the ethical competence of clinical managers, Erica Falkenström, aided by philosopher Marta Nussbaum, discusses the role of emotions in ethical dilemmas and decision-making situations. She discusses the way in which rational emotions differ from irrational ones; whereas the former are about the real-life situation and the real-life choices it demands, the latter are grounded in personal fears and prejudices, such as contextually unrelated feelings of inadequacy or anxiety. Woe betide us if we let gut feeling take over then.[17]

And likewise in the field of research it is about an educated gut feeling, one that comes from having profound knowledge of one's discipline rooted in many years of experience. Rather than any old emotion-driven whim.

As in the surgery, so in the lab. The deep, immediate familiarity-knowledge creates a purely physical sensation for the researcher as well as the surgeon. We feel it 'in the gut' that something is right or wrong. 'Where do you feel it when you discover something new?' I ask physicist Mattias Marklund. 'Well it's a gut feeling, isn't it?' he replies without a moment's hesitation.

It's a sense of euphoria which settles in your whole body. You get an endorphin rush, almost like that "runner's high" you can get when training – although you get it from coming up with a good idea. Everything goes so smoothly! It's interesting that you bring this up, because I think that what enables us to put up with all this drudgery, this donkeywork, is the knowledge that at some point we'll get that feeling, and that's worth *everything*.

'It's a kind of elation that resides in the breast,' exclaims political scientist Bo Rothstein, 'A feeling of "yes, bang, got it!".'. It's a real distinct physical sensation, a feeling of something dropping into place.' Unfortunately it's a rare pleasure, he smiles, 'and maybe I've already had my last taste of it.'

It is the same when something is wrong. That discomfort that the Gladwell statue experts expressed is well familiar to me. How a 'strange' result produced by a computer, one that lacks rhyme or reason and seems impossible to interpret, perches like a nasty feeling somewhere in the chest. 'It's terrible,' says Mattias Marklund, 'It really is the most horrible feeling there is, it almost makes it hard to breathe. It's hard to even talk about it.'

'It sits a little lower down in the stomach, a slight feeling of sickness,' says Bo Rothstein, describing what it feels like when he can't make head or tail of what he's seeing. When he finds no suitable explanation, sees no pattern and hits a dead end. 'I've been working hard all day without getting anywhere.' Trapped in the researcher's limbo.

As Peter Gärdenfors writes in *Den meningssökande människan* the body reacts with nausea when what we see clashes with the patterns we expect.[18] This very biological response seems to be deeply embedded in our evolutionary history, when the unexpected often spelt danger. The reactions of the savannah translated into unease in the presence of some arcane, meaningless figures on a screen.

What a relief when it later turns out to be a coding error, or when we suddenly hit on an interpretive framework that fits! How calm and warm that makes us feel. When the unease dissipates.

I would also point out in saying this that the embodied mental feeling I'm trying to capture has nothing to do with the field or method of study. Sometimes one might have the impression that this vivid, physical experience is the preserve of so-called qualitative research, by which I mean research based on prolonged close contact with the object of study and whose results are not expressed in numbers. But I say this is wrong. The intuitive, embodied, instant act of seeing is just as much part of quantitative research as it is of qualitative. We have I suppose just been better at convincing ourselves to the contrary.

I have never, for example, seen a quantitative researcher write the kind of text I'm producing here.

However, the physical euphoria that rows and columns of numbers can inspire in us, when they suddenly expose to us how something works, is, I'd wager, no less real than that triggered by a sudden insight half an hour into an interview. And how we actually organise our intellectual meeting places so that our bodies feel good and our minds consequently think better, is nothing that sets one kind of research method and tradition apart from another. We are all equal in the eyes of the body.

Perceiving Research Objects

The most important aspects of research we never teach. It has taken me a good while to realise that things really are this bad. But unfortunately it's true and has something to do with the perceived association so often drawn between the scientific and the unphysical, regulated.

In our methodology books, we can read about the process of social research. It starts with a theory about some social phenomenon. From this theory certain assumptions (sometimes called hypotheses) or questions are then drawn that can illuminate aspects or parts of this theory. Data relevant to the questions or hypotheses are then gathered or applied to already existing data material. Methods that are adapted to the data are then used to analyse the material and answer the original questions. Conclusions are then drawn from these analyses and their consequences for the theory presented.

It all sounds dandy. Organised and methodical. Clear stages. Only there is one little problem. Real researchers hardly ever work in this way.

Our questions come from all angles, sometimes from theories, but more often from an observation that piques our curiosity or from some issue of current affairs. Our data are rarely a snug fit for our questions, so we have to forever make compromises and interpretations using inadequate observations. We seldom master more than a handful of methods and sometimes try to stretch those we know to answer questions they were never intended for. Our conclusions are rarely based only on the observations we have made in data, but on the sum of our knowledge and our prejudices about what society is like.

Research rarely follows well-defined steps, or what in computer programming is known as (the entirely unrealistic) 'waterfall model', with its distinct stages and one-way flow. Instead, we skip and jump back and forth between vague ideas, theory fragments, analyses, data production and conclusions. We are guided by an intuitive feeling for what is important, one based on having seen similar 'situations' so many times before.

Instead there are two other processes that are fundamental to distinguishing between good and bad science. The first is about establishing research objects productively; the second about reducing complexity in a way that does not violate what we are studying.

Establishing a good research object is the key to successful research. It is here that the chaff is sifted from the wheat, it is this that most clearly distinguishes bad from decent or brilliant research. The successful constitution of the research object is that on which everything else hinges. From it flows theories, methods and strategies.

The establishment of a research object is about answering such questions as 'Why do/does I/someone want to know something about this?', 'What is this an example or case of? What larger class of events or processes or mechanisms is this part of?', 'What are the properties of this thing I want to study? What is it made up of?', 'What other things are similar to that in which I'm interested? What can I learn from this? What should I (not) do myself?'

The ideal research object possesses certain properties. One is that it points beyond itself – so that when we have learnt something about it, we have immediately learnt a good deal about something else. The knowledge it yields is transferable to other contexts and situations. A research object that is merely an example of itself is uninteresting; its lessons lead us nowhere, no horizons are broadened.

Another is that it generates researchable questions. Researchable in the sense that in it we can see what kind of information should be created, that this information actually can be created, and that the created information can be analysed using existing (or inventable) methods and approaches. If these three conditions are not met, the research object will remain sterile since no research activity can be derived from it. It might be a thought object, but not a research object.

Note that I do not say 'collect' information (as in the method books' 'data collection'). This can mislead us to think that the information is waiting out there somewhere for us to go and pick up – as if we were picking mushrooms or berries. But this is not how it works. We produce the information that we then analyse. This does not mean that we create reality at our whim. Not all information and not all data is there for the producing.

Even though I have just tried to describe what characterises a good research object, it is not something that lets itself be formulated in a few simple heuristics. No one can learn what a good research object is by reading about how to find a good research object. But we can learn to recognise one when we see one. Not by someone telling us how to do it but by time and time and time again being in its vicinity. Read, think, imitate, emulate. Laboriously learn to see. Develop a gut feeling. Ten thousand hours.

A good research object also creates peace of mind n the research group. When it is clear what is being done, what the problem is, how it is to be tackled. It might sound strange, but much of the work done at universities and in the world of research is tainted by an uncertainty about what to do and how to do it. A good deal of the destructive conflicts that plague many an institution and damage the research being conducted there are based on an uncertainty about the research object, the nub of all activity, the very purpose of the endeavour. These second- or third-degree disagreements are rarely productive; they just breed anxiety and animosity, lead nowhere. If you have a common research object, however, you have productive first-degree disagreements, ones about how a shared research problem is best approached.

Seeing a research object is more a matter of perception than thought. You feel it in the gut – in the whole body – when you see what is good, what is extraordinary. When the pieces all fall into place, the sudden insight, the inspiration. These are moments of unsullied joy – there is no other way to describe it. When you have the feeling of 'reaching to the brink of one's ability', as Sverker Sörlin puts it in his book on skiing (with the, in my context, apposite title *Kroppens geni* or 'The Genius of the Body').[19] And conversely, how bad the warped, the ill-considered, the clumsily formulated feels. The physical discomfiture is so intense that you can barely remain seated.

I am afraid that in this – this learning to see – many of the methodology books written and much of the methodology teaching done at our universities are more of a hindrance than a help. We are easily lured into thinking that good research rises automatically from a slavish observance of the rules. We easily have the impression that research is conducted in definite and distinct stages that follow a definite sequence. Research thus conducted is boring – correct but inconsequential. As though begotten by decree from the Ministry of Education. Research that stays on level three of the Dreyfus model: manufactured and executed by competent rule-followers, rather than virtuosos. Unembodied.

The same is true when it comes to the next most important stage of good research: reducing complexity. Here, too, the methodology books give no guidance. I do not even think I have ever seen this mentioned as an important consideration. In actual fact, it is absolutely fundamental. From the volume of information on the researcher's table can be extracted many stories. To create a coherent narrative from this hive of information, be it textual or numerical, we need to reduce the complexity of the data. Some information must be regarded as incidental, other must be sifted out as fundamental.

But how? We get some guidance from our theories. And there are many methodological techniques and computer programmes that can reduce complexity, whether the data is textual or numerical. However, the steerage the theories provide is often rather weak: there is always a considerable gap between our theories about mechanisms, processes and so on and the profusion of information we have in front of us. And the techniques are only useful if we know what we are looking for. There is not, nor will there ever be, a computer programme able to judge the newsworthiness and relevance of the material due for scrutiny.

Instead it is, again, about seeing, perceiving, what is interesting. Given how our research object is constituted, what in all this information is significant? This is where Malcolm Gladwell's *thin-slicing* comes in useful. To be able to see what is central and what is irrelevant on the basis of embodied, deeply experience-based knowledge.

Some people seem to have this ability in their blood. They are scientific *naturals* to use a boxing cliché. They have that eye, that slightly wicked analytical eye, that sees what is going on under the surface. Others can never learn. They should choose another way to make a living. But for most of us, it is the grind, the stamina, that gives us the expertise. To do, redo and redo again. To not give up, to constantly strive to stay within reach. Until suddenly we have it.

Or in the more poetic words of author Per Olov Enquist: 'But that which does not connect, that gives insights of the kind that are the most difficult and the hardest to grasp, that is no river, it does not lead inexorably on from one point to the next. It must be watched, intently, until suddenly.'[20]

Learning to reduce complexity is therefore done by example. By seeing what others have done, by carefully observing, imitating, adapting. Author Erica Falkenström tells of how she watches her father, artist Stig Sundin (1922–1990), meticulously study the legendary Gotlandic stone-cutter John Larsson (1897–1986).[21] To see how he held his tools, how he formed the stone under his hands. Virtually tirelessly, hour after hour. To learn. It is thus that the researcher learns from her forerunners. If she has understood that this is what needs to be done. If she has the energy and stamina.

I have been to the Rock Museum in Kettelvik on the island of Gotland, where John Larsson's tools are kept, and where you can see a greatly enlarged photo of the man himself. For me, this image is the very emblem of deeply embodied knowledge: how the tools rest gently in his large hands, as if they are growing out of his body. No gestures, fancy moves or pedestals, just the embodied calm that comes from full mastery of a craft. The researcher who can analyse like John Larsson cut stone has reached the furthest limits of expertise.

The last aspect of good research is about seeing and creating implications. The stories that have their genesis through the establishment of the research object, the production of data and the reduction of complexity must be reconciled with the collection of theories, assumptions, presuppositions and research strategies from whence they came; only then can we see the contribution of what we have produced. It is a rather hairy moment: what if all this toil has been in vain? But since we have already anticipated this possibility during the earlier stages of our research and have already faced it in our minds several times, we have had time to adjust our course and our point of emphasis along the way. Provided, of course, that we have understood that this is how research works.

And going forward – how can we learn, and teach? I would like us to start by talking to our students. Not give them the polished, bowdlerised version but the raw, X-rated one. The one where our own weaknesses and shortcomings are exposed for all to see, but where we still dare to claim that we know what to do.

This is particularly important in the encounter between scientific knowledge and practical skills, something that has become increasingly common with the 'academisation' of teacher and nursing training programmes. If it is the polished, regulated, method-book version that serves to represent academic know-how, there is a clear risk that it will 'colonise' the more practical, concrete and physical forms of knowledge.

This leads to a downgrading of experiential and situational knowledge, something that Ingela Josefsson writes insightfully about in her books on professional know-how and forms of knowledge.[22] Ultimately, deep professional know-how is in jeopardy. But if it is research-based knowledge that understands its own genesis and character, opportunities are created for a more open meeting of different forms of embodied knowledge, one in which both sides can genuinely learn from each other.

Sverker Sörlin's *Kroppens geni* tells of the conflicts between different training philosophies in the field of Norwegian skiing.[23] Of how volume-training, the 'relaxed long run' that has been the traditional basis of the Norwegian training model, has been gradually challenged by other faster and more patently scientifically based methods. And of how these different forms of knowledge have complemented each other and thus been made to develop; but where the training methods that 'just' build upon deep practical experience have never been beaten by those that derive from controlled scientific experiment. Of how no one really knows why the 'relaxed long run' produces such miracles, only that it seems to work. As Sörlin writes about the world's best skier Petter

Northug: 'I suggested an uncertainty in the knowledge base. He wants me to know that for him there is no uncertainty. He has asked his body so many times that he knows the answer. And he trusts it.' The credo of embodied knowledge could not be formulated any more clearly.

Therefore we must make our students learn by example, not through lectures. As creativity researcher Teresa Amabile puts it, one of the most important contributions the social environment can make to creativity is in the models it presents.[24] Do it like this – or tweak what is done. This is how things work. In the real world.

By seeing how problems are solved practically, how one de facto acquires knowledge about reality, one can slowly learn what to do. Through the power of accreted example. Now, someone might protest that this is exactly how the best methodology teaching is already done, that I'm kicking in open doors. Maybe, in the best cases. But established practice is still, unfortunately, to teach rules of procedure from the lifeless soul of the methodology books, so that the knowledge gained can easily be measured in standardised form.

However, methodology books are still needed in one form or another, so it is important that they talk about how research actually works instead of conjuring up a what-if world. I would love to see a methodology book whose four chapters were called 'Constituting the research object', 'Producing data', 'Reducing complexity' and 'Creating implications'. A book that systematises what we actually do when we research. Maybe I should try to write it myself. But first I have another story to tell.

The Good Space

At breakfast on the last day in Cadenabbia, Edeltraud leans forward: 'I must ask you something. We've been sitting here for three days and complained and complained about everything you've done. But you're still happy and attentive and reflective. How is that possible?'

Just as I open my mouth to answer, Erik interrupts: 'Because they care about *the truth*, not about their careers!'

We laugh. The day before I'd asked Erik – they'd actually just been complaining for three whole days – if we'd ever manage to put a book together at all. He looked at me and replied with his usual genial ruthlessness: 'Well, now I've got to know you and know how dedicated you are, I'm sure things will work out well in the end. But I can tell you that if someone had sent me *this*' – waving his hand vaguely in the direction of our piles of paper – 'from a publisher's I'd have said, "No, no, don't touch it with a bargepole."'

The strange thing was that it felt perfectly OK.

We finally did manage to produce a book. To crown it all, one 'that sets a new standard for future analyses in political sociology' as a (arguably overly) gushing review was to claim.[25] We revised some of the book's chapters from scratch, we edited and fixed and cleaned up. It turned out really good in the end.

But Erik was wrong about why we were still upbeat after having been the target of three days of incessant intelligent criticism. It wasn't because we didn't care about our careers or because we had some kind of extraordinary longing for the truth that we were still in a good mood. It was because we felt secure with each other.

A few years earlier, at the project's very first meeting, I related a story I'd heard from the American sociologist and historian Rogers Hollingsworth,[26] who has spent many years trying to understand what it is that makes some research environments more successful than others. In one of the many interviews he has held to this end, a research leader says,

> It's a matter of climate creation. Think about when you show your project colleagues something you've written and they don't think it's any good. It doesn't bear up, they say, and you'll have to start again. You go back to your office, bin what you've done and start from square one. Then you come back with something new. But they still don't like it and say you'll have to do it all over again – again. Do you get angry then and call them idiots, or lose heart and blame your own stupidity? Or do you go back to your office, scrap everything and start over with the intention to present the third totally revised version to them? That is what set the really good research environments apart from those that are simply OK.

A ripple of nervous laughter around the table. Ooh, I wonder what's coming, joked Staffan. But I explained that what I was trying to say was that around our table there should be honest and selfless conversations about our research, respect for others and a willingness to take their efforts seriously. That criticism is something to welcome, that it means someone has bothered to care so much that they've gone to the trouble to find flaws in one's work.

Sometimes, perhaps, I think that I was too successful at creating a cheerily disrespectful atmosphere. During the years we worked together I would some-times have my texts described as the 'usual laundry list' and hear again and again that if I just did things differently it would doubtlessly all turn out well. But project members who'd rather continue discussing my book manuscript than taking the 'more sightseeing in Florence' option you can't help but love.

My project would never have worked like this if we hadn't liked each other. It is only in relation to people we like and trust that we dare show our vulnerabilities. It is only here that we dare to say that we don't understand, that things didn't quite turn out as we'd planned and that we need help. Even though things almost never turn out as well as we want and even though we almost always need help.

It's about creating (figuratively and literally) spaces where we feel comfortable. Only in such spaces can the group's dynamic be positive, can people feel so at ease that they remember all they know. What happens in these spaces is something almost magical, something that could never happen in the head of one of the individual researchers if they had sat on their own.

Before getting down to our project I'd always been somewhat sceptical about the idea that the research group was to be something more and greater than the sum of its members. This was also a thought I voiced when we planned the application that was to serve as the basis of the project: we'll apply for joint grants for things we intend to work on individually but under a common umbrella – which is exactly what many applications de facto look like. No, said Maria in her stable, candid manner. If there's to be any point in our applying for grants, we'll actually work together rather than just pretend to do so. OK, I replied a little uncertainly, but soon allowed myself to be convinced. And that's something I'm very grateful for.

This is because a well-functioning group has what are referred to as emergent properties. This means that at a group level, new qualities emerge that cannot be reduced to any one of its component parts. Something I as a sociologist ought to have taken as understood, one might think.

Creativity researcher Ken Robinson talks about the 'alchemy of synergy' which arises when we get to work in this way with like-minded yet different people.[27] About how by 'finding our tribe' we can create a context in which the whole is actually greater than the parts. He cites philosopher of science Michael Polanyi on the significance of environment, on how it has to grow from out of a common interest, and how attempts to forcibly summon a creative research environment are doomed to fail.

Robinson argues that the best teams have three defining properties. They are *diversified*: composed of people with different yet complementary abilities. They are *dynamic:* characterised by a high degree of interaction and communication and the ability to give and receive criticism. They are *distinct*: assembled to solve specific problems or work with particular tasks. When they're done, they move on to something else.

The group that under the direction of Miles Davis created the immortal *Kind of Blue* is an extreme example of this. The group had been formed only a few months before but, over two recording days in the spring of 1959, arguably the most famous album in jazz history was born. Without preceding rehearsals, in the inspiration and improvisation of the moment. But the alchemy of synergy occurs every day of the year when creative people manage to cooperate and do what none of them could have managed alone.

In our project, things happened, in those best of moments, that astounded me. When suddenly someone says something that fills a vital gap in my argument or that makes me see glaring errors of logic. That were there but invisible to me. When it's like we're talking with one voice so that afterwards it's difficult to untangle who contributed what. A collective intellectual force that is greater than the sum of the individual members' minds.

One of my colleagues described the virtually opposite situation, in which he tried but failed to establish a similar forum. 'My simple thought was that this would be an occasion when we could present what we were wrestling with at the moment, ideas that we couldn't quite formulate, analyses that we needed help with, where we could ask for advice or tips,' he tells me over lunch, 'so that we could get help from each other and so that we didn't have to sit alone in our offices tackling our own problems ourselves.'

But it never worked.

Everyone was so timid and guarded. The doctoral students were afraid of appearing stupid and ignorant, and the established researchers were so competitive with each other that they weren't able to reveal their weaknesses. No one really said anything, even though I stressed that this wasn't a place where anyone had to prove how clever they were. After a while I ditched the whole thing in, it was all a total waste of time.

No trust and security, no cooperation.

Rogers Hollingsworth has scrutinised the particularly excellent research environments, those that turn out one Nobel laureate and groundbreaking discovery after another. Places like the Rockefeller University in New York, or the California Institute of Technology (Caltech).[28]

There are two factors that Hollingsworth deems particularly crucial. The first is just the right amount of differences in the group and at the institution; Goldilocks diversification if you will. Too little leads to stagnation. If everyone comes from the same discipline, has read the same books, attended the same courses, applies the same methods, they have little to bring to the common table. If you know exactly what I know, there is nothing I can learn from you and our collaboration is pointless.

When I meet political scientist Bo Rothstein to discuss the organisation of research, he identifies 'complementary competencies' and 'collegial competition' as key. Admittedly, at the Quality of Government Institute, which he has established, most of the researchers are political scientists. But they have slightly different areas of expertise: one is a specialist in a geographical area, another is driven by specific theories and a third uses a particular method. In one sense, all these researchers compete with each other. For future positions, attention, reputation. But in another, they collaborate and help to elevate each other. As co-authors, as reciprocal problem-solvers. The entire group benefits from this, and the idea that we *either* collaborate *or* compete is fallacious, argues Rothstein. In most cases, both aspects are present in more or less large doses. In which case, if their collaboration is to be more than the sum of its parts, the researchers cannot be mere clones of each other.

Differences are important, not least for offsetting the tendency towards over-specialisation that forever lurks within science. The way that the doctor who has done research on the knee ends up knowing nothing else and has trouble seeing how its condition is related to the rest of the body. Or the political scientist who has difficulty seeing how the subject of their niche erudition relates to other institutions and organisations elsewhere in the body politic. In such cases, colleagues and their differences are essential to the upholding of perspectives and opening of insights.

But if the differences are great, it will be impossible to understand what another is saying when she tries to explain her work and ideas. We might not even have the same sense of purpose. In such cases, cooperation is impossible.

Rothstein describes a collaboration with lawyers that largely failed because the disciplines were so different. With the causal logic of political science ('how can this be explained') up against the subsumption logic of law ('what text is applicable in this situation'), neither side drew much charge from the synergy. 'They were lovely people but our collaboration was fruitless. Our mindsets were simply too different.'

So, Goldilocks diversity is what we want – not too much, not too little. With maximal communication, which is the second central factor identified by Hollingsworth. There is no point sitting down the same corridor as the world's brightest researchers with knowledge superior to yours if you don't talk to them. Therefore, the most successful research environments are those that have found ways to maximise the communication between researchers. And it is about creating spaces, physical as well as intellectual, where these meetings can take place.

At the Rockefeller University, this was manifested, for instance, in an apparently bizarre concern about the actual layout of the physical environment. The lunch tables were to have exactly eight seats each, eight being the maximum number of people, it was argued, in which a single conversation could be sustained without the group splitting into subgroups. Fewer, and input would not be maximised.

But they also arranged interdisciplinary seminars, where the physicists had to read the chemists' essays and then report back to the chemists what they got out of the texts. And vice versa, this way and that. By their constantly having to think outside their own frameworks, their space was expanded. They could see analogies, parallels and other ways to solve problems.

If everything is to work, you naturally need a good helping of courage and self-security. Dismounting from your scholarly high horse to become a novice again; daring to admit that you cannot do it, you do not know it, you need the others. An intellectual leadership able to cultivate a secure environment that can entice people to step outside their boxes is therefore also necessary. Where the genuinely new can be found. Unfortunately, I have no answer to how this intellectual leadership itself can be cultivated, or why so many outstanding researchers do not seem to realise its importance. What I do know is that if something is to get better, we must start by acknowledging its value.

Another interesting finding that Hollingsworth makes is that the ground-breaking environments are rarely the most productive. There are other places that produce *more* research, but not that produce *better* research. This is probably because it is quicker to publish yet another article or book that is little different to many others that have gone before than it is to come up with something genuinely pioneering. This is worth bearing in mind, especially because our current ranking and resource-allocation systems prioritise quantity over quality.

Organising knowledge production is very much about organising meeting spaces. And as we have seen, the physical conditions are significant. Many institutions – especially in the United States – lack places for researchers to meet. There is barely a coffee machine, let alone a lounge. They are all friendly and charming once you meet them, but you have to arrange a time. Which takes already knowing that you *should* meet. And then the risk is that you only ever meet your academic clones.

Many years ago, as a young doctoral student, I heard perhaps the most famous Swedish sociologist declare over dinner that he had finally got his colleagues at his new Dutch department to understand that 'they were to sit at home and write instead of plotting at the department'. I thought it sounded a little strange – at my department we generally sat working as we had coffee. (After having understood the climate at the Swedish department he had left behind I sympathise a little more.) But as he was a prominent professor, I thought there must be something in what he was saying. With time, I have realised how fundamentally erroneous this reasoning was, how destructive for a research environment it is when ideas like that gain purchase. Without regular collaboration and meetings the intellectual flow stagnates.

Political scientist Li Bennich-Björkman's study of innovative and stagnant research environments puts special emphasis on the importance of meeting places.[29] People actually occupying the environment rather than sitting at home is an absolute prerequisite, but far from enough. Dedicated physical meetings spaces in this environment are also needed. A lounge for the more informal meetings, a well-appointed seminar room for the more formal ones. Such meetings are essential to both communication/information as regards what the respective parties are doing and to building a spirit of trust and sympathy.

However, for these meetings to be productive, Bennich-Björkman holds that there has to be a common research agenda and an academic leadership able to cultivate a sense of shared interests. If I am completely uninterested in what you are doing, we will probably speak of other things than research when we meet at the coffee machine; if we are also rivals and dislike each other, we will avoid each other. But if we have the same thoughts and opinions about everything, we will derive nothing from this encounter. Because I knew everything you know even before we met.

Bennich-Björkman writes of 'that subtle web of social relations which make up the determining cultures that lie at the very heart of the university'.[30] This culture is what determines whether people feel they can trust each other, share problems and insights, and collaborate. This trust, communication and collaboration decides, in turn, the quality of the knowledge production. And this culture, she continues, is difficult to control or manipulate from outside or above. It just emerges from encounters in the everyday knowledge-producing spaces.

For many leading researchers, the group or the team is such a natural reference point that they are almost blind to any other options. When I meet memory researcher Lars Nyberg to discuss the everyday organisation of research, it is all but impossible to make him talk about himself as an individual researcher, or even use the word 'I'. Even though he is probably my university's brightest research star with a global reputation and a publishing record that most researchers can only envy, he consistently uses the word 'we'. How they work as a team towards a common goal, in competition with other teams elsewhere. About the joy of working together, about the group succeeding in its endeavours. If there is one aspect of all this fun that he finds hard, it is having responsibility for the team: that its survival and the livelihood of its individual members rest very much on his, and only his, shoulders. That everything is so fragile.

In Christer Sandahl et al.'s research on the 'strategic research programmes' at Lund University the central role of the group is confirmed.[31] Regardless of whether one asks about the structure of the work, inspiration, identification or mentorship, the research group leader is always at the heart of things. Roughly half of those asked think that the research group leader has such functions 'to a great extent', another 40 per cent feel that it is to 'to some extent'. As for the formal managers – the heads of department – only a small percentage think that they serve to a great extent as a source of inspiration and identification, and the situation is not much better for those who coordinate the large research environments. It is in the small contexts of everyday life and intimate meetings that inspiration and identification arise. Already at the departmental and environmental level the bonds are quite weak and contacts cool and formal. Sandahl's studies ask no questions about the 'university management' but I'd wager that few people if any would say that they drew their inspiration and identity from there.

The world-leading psychologist Gerd Gigerenzer leads a division of the Max Planck Institute in Berlin. In the preface to one of his books, Gigerenzer writes of his experiences organising research in a multidisciplinary environment, claiming that '[t]he single most important success factor is setting up the right environment'.[32] Time must be given for everyone to have a coffee break together, not because it's obligatory but because it's pleasant. There has to be a seminar at which the participants hold discussions that cross subject and disciplinary boundaries. And everyone must work on the same floor. When members of a department work on different floors, writes Gigerenzer, collaboration decreases by 50 per cent; this, he goes on, is because since our early evolution on the savannah, we have orientated ourselves horizontally. The truth of this theory notwithstanding, the organisation of meeting places is a matter of life and death for a research environment. It can be the coffee lounge, it can be the common dining room, it can be the seminar room. Easy to see, hard to create.

Similar ideas informed the design of a new building at the Australian National University (completed in 2007), where the once dispersed political scientists were to congregate.[33] I meet Chris Reus-Smit, professor of international relations, and the man behind how the premises were to be laid out and occupied by departments and researchers. Chris says that his idea was to realise several things with the new premises: to create meeting places where people could more or less randomly bump into each other; to retain the integrity of the formerly separate departments while creating new collaborations and meetings across departmental boundaries; and to reduce the hierarchical gap between professors, senior lecturers and doctoral students. 'By organising the physical room, you build social relations, identities and fellowships,' says Chris. 'We wanted to get away from the dead corridors with their closed doors, and to facilitate visibility and transparency and interaction.'

The new building gathers the meeting places in its middle around a large atrium. Along the external walls are the offices, fitted with glass doors to contrive a sense of light and visibility. 'I'd say it's the only part of the plan that hasn't at all worked,' says Chris. 'Pretty much everyone instantly pulled the blinds down over their doors. That visible the researchers clearly didn't want to be.'

Each department is 'staggered' on the floor with the next, so that people from one are constantly encountering people from the other in the corridors. To ensure that the departments still remain connected, they have placed the departmental library on one floor and the post room on the other. 'These are the two places everyone goes to, so we can guarantee that the researchers move up and down the floors,' continues Chris.

Different hierarchical levels are also mixed within each department, so that there are no 'professor' or 'doctoral student' corridors. Chris has also had all titles removed from the nameplates, so that a visitor to the building has no idea if they are entering a room occupied by a professor or by a doctoral student. The internal and external message is clear: Here it's the force of your argument that counts, not who you are in a formal hierarchy.

'But it wasn't easy to get the budget department and the university management to go along with all this,' Chris recalls. 'I had to argue until I was blue in the face that it was vital to academic quality, that we'd produce better research and teaching if we organised our rooms like this. That even if it cost more money, it would pay off in the long run.'

The general opinion is that the whole thing was a success. The interaction within and between departments was greatly enhanced and those fortunate enough to be on the premises can't understand why all departments aren't organised along similar lines.

When we met, Chris Reus-Smit had his office in the former monastery of Badia Fiesolana outside Florence, which houses the European University Institute. We hold our conversation in the refectory, where everyone working there goes for their lunch. 'So what about this building?' I ask apropos of what we have just been discussing. 'Here, the researchers can barely fund each other's offices. If I disappeared it'd take weeks for anyone to notice I wasn't here.' 'It's lucky we have the refectory,' he says. 'It's virtually the only place we have here to meet.' And that is no lie; the previous week I had, quite by chance, heard about his experiences in organising physical spaces for intellectual meetings – across the lunch table in a conversation that was originally about something quite different.

What all these attempts to influence the physical environment have in common, be it the Rockefeller University dining table, Gerd Gigerenzer's coffee meetings or Chris Reus-Smit's mixed floor, is that they establish conditions for communication and impromptu interaction. To quote Malcolm Gladwell again, they 'created the conditions for successful spontaneity'.[34] Success in our context meaning the contrivance of new explanations, new interpretations, new opportunities to examine.

For one can see creativity and innovation as *states that are essentially by-products*, to borrow from political philosopher Jon Elster.[35] One cannot be innovative or creative by trying to be innovative and creative. Just as little as you can be happy by trying to be happy, or impress by trying to impress. Innovation and creativity emerge as by-products of being actively curious in a real problem-solving situation. And here is where spontaneous meetings are essential to the broadening of horizons, the prevention of tedium and the discovery of new possibilities.

Organised spontaneity might sound oxymoronic, but the point is that it is the *conditions* for spontaneity that can be organised, and this through the creation of meeting places. What people then do with these spontaneous meetings is another matter. One just has to trust that intelligent and creative people, when placed in the same space, will integrate – and if they do, that something new will arise. From the good meeting springs knowledge.

But not all meetings are good and not all teams are effective. Groups are sometimes rife with rivalry, envy and petty cruelty, and people only work with colleagues they like and trust and who like and trust them in return. Ad hoc teams hastily formed for the sake of obtaining a grant are responsible for many wasted years and ruined careers. Failed collaborations can be hell on earth.

The Bad Space

Let us call her K. This is the story about how she quit the world of research. About how she and research parted ways.

It was time for the year's subject conference. K was a relatively fresh doctoral student, but that didn't stop the prominent professor from laying into her research. The kind of research she was doing had no right to exist in the discipline, he declared, no one should be spending time on such matters. 'I felt really lousy,' K says. 'The most enduring feeling was of enormous shame.' At being a poor researcher, unable to defend herself, not good enough. A feeling of being worthless. She 'almost fell into a daze', couldn't really hear what was being said, staggered back to her hotel room. The next day she could hardly drag herself out of bed.

A month's sick leave later she was back, but her research was never the same again. She never recovered her passion, no longer spent that much time at her department, accompanied her partner on his postdoctoral work abroad, became pregnant and went on parental leave, got a job outside the university, and gradually removed herself body and soul from the world of research. It was a long process, but in actual fact it ended that day it all fell apart. 'It was a bad space,' she claims in alluding to my spatial associations. 'One in which I was truly crippled.'

It is likely that the flak she received would not have been so devastating if it had not augmented a profound insecurity about what one is actually supposed to be doing as a researcher. What was the core of the craft, what knowledge and skills should one have, what tools did one have at one's disposal? And then these ever-present conflicts, between schools, between individual researchers.

To be sure, she received the support of her supervisor and the other professors at her department, as well as of her fellow doctoral students, after the seminar. 'But it wasn't the kind of support I needed,' she continues. 'They were so involved in the battles raging in the subject. They were all us against them. But I wanted none of it, didn't want to fight. I just wanted to feel that I was OK, that I was entitled to be there too.' She couldn't really talk to anyone about what had happened, and after a time she had repressed so much that she even found it hard to conjure up the memory. A deeply unpleasant sensation was all that remained.

There's no need to feel sorry for K. She has a good job, in a meaningful context, with a better salary and more stimulating work that she would have had if she had stayed put. 'Where I am now I don't feel like I'm constantly being judged, like I did at the department,' she says. 'Here it's more about working with others to get the job done, about how we can be better. It's not me as a person that's at stake.'

It was probably the best outcome for K. But does the world of research really have the right to treat people like this, and can it afford to? Should they not be made best possible use of? Surely we should not poison and ruin the environments for which we ourselves are responsible? How can we argue in favour of independent research if this just leads to power games and oppression?

I haven't only experienced some of my best moments as a researcher working with others, but also some of the worst. Moments that have made me wonder if this is really where I want to be, if I too should not quit this toxic space.

I remember, for example, a seminar at one of our larger universities. I was some way through my doctoral education, had written a critique of the use of *rational choice* theories in politico-sociological analyses and was given the opportunity to present the paper at a seminar. My work apparently was not to the taste of the professor leading the seminar. His questions seemed ostensibly innocent, delivered in a low voice, and a transcription of the conversation would probably have given no indication of unpleasantness. But something threatening lurked in the questions, so that to my ears his sibilant voice sounded more like a hiss. He was certainly out to get me, to crush me even. And under the table I could see his agitated, training-shoe-clad feet. How they incessantly wound around each other, revealing that he was a bundle of pent-up aggression. And this was not just my overstrung imagination talking to me. I could see the physical discomfort our argument was causing around the table, how the other seminar participants were itching to get up and leave, to get out, to go home.

But let us not speak ill of the dead. And if it were so that this was an isolated nightmare example and a relatively restrained one to boot, it would not be worth mentioning. (He never actually *said* that I was stupid, just made it clear that he thought so.) But these bad meetings take place every day, at universities around the world, in seminar rooms where dreams are shattered and self-esteem is pulverised. It should not be this way.

Sometimes I have been there myself, on the wrong side. Sometimes my comments have been driven by a desire to wound, quash, silence. Sometimes my contributions have been based on wanting to appear smart at another's expense. The thin membrane between care and cruelty is so easily crossed.[36] I am not proud of what I have done. But seeing one's own role in destructive situations is still the first step towards being able to change them.

It is about feeling uncomfortable and unwelcome. Especially vulnerable are those who for one reason or another are further removed from the ivory towers of academia than others: women, the working classes, immigrants. Those without academic traditions at home and in their immediate environments. Those with the wrong gender, wrong background, wrong skin colour. But this does not mean that others cannot also be affected. The white heterosexual middle-class man can have his self-esteem wrecked and his self-confidence obliterated when a sufficient amount of negative energy is trained on him. The bad contexts I am referring to here are just as much a barrier to knowledge as even the least ad hoc research policy.

Maybe some of these bad spaces have arisen because some researchers have misunderstood the role of the team in knowledge production. I talk to management researcher Christer Sandahl, who tells me that some research leaders who attend his lectures believe that their role is to whip the team into success, to steer them in the appropriate direction with an iron fist. The myth of the brilliant but heavy-handed researcher, feared but admired by his subordinates, is a hard one to dispel. Even though everyone ought to see that it doesn't work. 'Scared people aren't creative,' as Bo Rothstein notes laconically when explaining why 'academic anal sadism' is not only unpleasant but also destructive to research. If the unpleasantness could buy research success, it could possibly be endured and excused. But broken eggs do not always an omelette make.

Someone who has taken these unpleasant experiences seriously is the Swedish-Norwegian sociologist Karin Widerberg, whose book *Kunskapens kön* portrays the sense of unpleasantness and distance that women in academia can experience.[37] With powerful, vivid examples from her own life in research, she shows how women can be subordinated, belittled and ignored in the spaces that ought to be the forums of free exchange and open argument.

However, I am not so convinced as Widerberg that all this is a question of gender. That it is always men who are the oppressors and women who are the oppressed. Many of the examples of oppression, unpleasantness and distance she presents I can relate to myself, can feel in my bones. Am I then to infer that I am a subjugated woman? If not, it must mean that there are other dimensions than gender in which such experiences arise. And this Widerberg must surely realise, yet on she goes with her facile account of male superiority and female inferiority.

This causes a peculiar blindness to all the other dimensions in which power and hierarchy are engendered and reproduced in academia. Perhaps it is from somewhere here that the self-evident way in which some female professors lay into their younger fellow-females comes. That they simply do not see what they are doing, that they so naturally see themselves as the *underdog* that they do not notice when they themselves are the oppressor.

Women and men in academia are hardly homogeneous collectives with shared experiences, interests and values. And yet this is how they are depicted in Widerberg's book. The result is the strengthening rather than the weakening of the gendered, when the latter should be the aim of every gender-critical analysis and practice. Gender is essentialised and other experiences are subjugated.

But her observations of the body and its discomfort in research I take away with me. The nasty gut feeling that there are people out to harm you, that you are not welcome, that you do not belong in these rooms. The creeping feeling of having to escape lest your mind, your soul, becomes irreparably damaged.

Research collaboration is not always pleasant. Sometimes it is everything but. Sometimes 1+1 does not equal 3 but 0.5, if that. In such cases, it would have been better for the researchers to have stayed in their rooms doing their own thing. When you cannot agree, when you fight with and are even cruel to each other. Mutual like and trust is fundamental to effective collaboration.

There are two particularly effective ways of destroying the research environment. I have seen both in action, many times. I call them pedestal construction and pedantic destruction.

Pedestal construction is about raising the person whose products are to be discussed to such heights that critical conversation becomes difficult, if not impossible. Like when the seminar leader spends not a minute or two, but ten on opening the meeting with a catalogue of all the publications, invited lectures, awards, academic membership and God knows what other feathers the day's presenter has in his or her cap. The pedestal is built higher and higher so that in the end you can hardly see the fallible human up among the clouds. And the desire for an open and critical exchange quickly shrivels to nothing.

Or the seminar commentator who opened by saying that we 'all should be grateful to D and T who have given us this'. Grateful. As if they were gods bestowing their gifts upon us. Who wants to argue with divinities? Not me, at least. So I sat silently among the audience.

And like all idolisations, they risk striking back at the vaunted without warning; he or she proves then to be nothing but human, maybe even flawed in their calculation or understanding – and, all of a sudden, the pedestal becomes a coffin.

While pedantic destruction is in some respects the opposite of pedestal construction, it almost has the same effect. This method is about fault-finding, about reducing the intercourse to a kind of mud-wrestling match, the goal of which is to find omissions, miscalculations, defective models, careless citations or erroneous conclusions – without first asking what we have learnt, what the bigger question is we are trying to illuminate, what this is a case of. Meticulousness and care are research virtues, but when they are the *only* virtues tedium sets in as they are transformed into pedantry and timidity and *doing things right* becomes more important than *doing the right things*. The researcher and the research shrink: ever greater trivialities are treated with ever greater precision and thought.

Both the pedestal builder and the pedant spread anxiety. Anxiety about making mistakes, making a fool of oneself, asking stupid questions, exposing oneself and one's inadequacies. The group's dynamic turns bad, confines us, locks us in. And the room around us shrinks and shrinks until in the end we can barely breathe.

A Big Imposter

Hand on heart, who actually likes being criticised? Don't we really just want praise? To be rubbed up the right way, esteemed, acknowledged by people we respect and like? It isn't nice to have our weaknesses exposed, especially to an audience, and particularly especially if we realise that our critic is right.

And yet we must by necessity expose ourselves to criticism if we wish to grow. As the self-help literature puts it: the inability to endure acute pain makes for chronic suffering.[38] Receiving criticism is the acute pain for the knowledge creator, and if we cannot tolerate it we can expect chronic suffering: stagnation, melancholy, pointlessness.

Or to use the language of counterproductivity theory: the right act is that which strives for long-term goals, values and interests, albeit at the cost of short-term losses and suffering.[39] The long-term goal: that we and the text we produce must be better. The short-term loss: we are painfully made to realise our own inadequacy and incompetence.

Being able to accept and use criticism creatively in a way that improves what you are creating and that helps you grow is the be all and end all of your research communication. But it is not easy, for criticism can easily awaken one of the greatest bugbears of intellectual activity: to be exposed as an imposter.

I myself am dogged by a pair of demons. Usually they are dormant, but they have been known to come to life. Particularly if things have not been going well for a while with my writing and publication. One whispers: 'It's been going well so far. But it won't be long before they find you out!' The other hisses: 'You did some good things once. But what have you done recently?' They are nasty little companions and I am unable to rid myself of them. They are silent for long periods, when my research is going well and my self-confidence and self-esteem are high. But as soon as something starts to sag and totter, there they are again.

American sociologist Dalton Conley argues that fraud anxiety, the fear of being exposed as an imposter, is actually endemic in our walk of life.[40] We create, just like in the world of advertising or culture, nothing tangible. We produce texts, not tables and chairs and kitchen drains. We forge ideas, visions, issues. And lacking materiality, they acquire meaning only in the minds and practices of others. In such a business, posits Conley, fraud anxiety constantly lurks. What if the result of my efforts lacks value, what if it is trivial, what if no one believes what I say, what if no one wants me? What if I'm just one big imposter?

What makes this fraud anxiety so fiendish is that it will not be muted by acknowledgements and awards. Because, you question, do you really deserve all these rewards? It is easy to feel like the malcontent in one of Jan Stenmark's drawings: 'My renown was now so great that I felt like a little shit in comparison.' As the pedestal rises, so rises one's own standards for what constitutes an acceptable performance, and with them, one's own performance anxiety. *It's been going well so far, but it won't be long before they find you out.*

If the fear of being exposed as an imposter is the one major internal threat to drive and creativity, acedia is the other. According to Wikipedia, acedia is 'a state of listlessness or torpor, of not caring or not being concerned with one's position or condition in the world' that is generally considered something particularly common to 'monks and other ascetics who maintained a solitary life'. In the field of science, acedia is defined as a 'gradual withdrawal of motivation for research and an increasing alienation from science'.[41]

Acedia is thus not synonymous with the little pricks of ennui that can deflate us all during the course of the working day. No, this is a much more profound feeling that all we are doing in our research is bereft of meaning, that our deeds are nothing but futile. Or maybe even that one's whole intellectual being has lost its meaning. That what we are writing and thinking is without worth and substance. Acedia is a dangerous condition indeed – it can end a career if it is not cured or at least alleviated.

In an insightful article from the mid-1960s, sociologist Hans Zetterberg describes the phenomenon,[42] and identifies its different sources: sometimes acedia comes from overspecialisation, sometimes from an inability to concentrate; it can come just as much from a failure to achieve one's objective as it can from overly sudden success. It emerges slowly, like a creeping sensation that something is not right, that what one is doing lacks meaning, that one's accomplishments are inferior and in vain. It grows worse with time if measures are not taken to parry it: every unproductive day increases the feeling of pointlessness and failure.

His description is so vivid that one might suspect him of having had firsthand experience of this acedia.

Yet I am not particularly convinced of Zetterberg's proposed solution: that extraneous reward (e.g. income) and the status symbols of the scientific system and society must be more clearly geared to scholarly achievement. More and better publications must garner higher pay and more distinctions and appointments, is what he is essentially saying.

But can a higher salary pegged to our publications really save us from acedia? And do we really need more honorary doctorates and academic fellowships? I find it hard to see how this would work. Acedia is an experience of the soul, an existential problem in which our humanity meets an endeavour that for us has lost all substance. Can 5,000 more a month and a medal on our chest remedy this? Permit me to doubt.

So what will save us? I believe that we can seek rescue – or at least respite – from both acedia and fraud anxiety from the social and emotional relationships that we create and sustain in our everyday work. It is in these relationships that meaning is generated and preserved, it is in and through them that fraud anxiety can be kept at bay. It is about someone genuinely caring about what we are doing, someone who is honest and spirited enough to tell us if we have started to ramble or ended up on the wrong track.

In this sense criticism – in both directions – is, well, critical. Not in a sense of attacking, dominating or crushing the other, but of taking the other and what he or she is doing seriously. Physicist Mattias Marklund argues that the local environment and the criticism one receives in it is absolutely fundamental. To be sure, we have all this *peer review* system and its inherent critical function, says Mattias, but it is still in the local environment 'where we know no one means us harm' that the most frank and constructive comments can be voiced. Because criticism is not always negative; it is often, instead, a matter of 'learning to be satisfied' and 'not sacrificing too much' in the battle against fraud anxiety. Of receiving acknowledgement that what you have done is good enough, that no more is needed (just yet).

I myself often think and talk about this as 'avoiding the nerve coat'. During my brief and undistinguished time as a cross-country skiing trainer, I learnt the meaning of the phrase nerve coat. This is when you have perfectly waxed skis but start to think during training that other people's skis seem to have a smoother glide. Should not the waxing be improved a little before start? What all's said and done? And so you get to work on the already perfectly waxed skis at the last second. Adding the nerve coat. And ruining everything.

Researchers also have to avoid the nerve coat, the eleventh-hour revision that just makes the text worse. The revision that comes from not really believing that your work is good enough, from the wavering of your self-confidence or self-esteem.

In his memoir *Ett annat liv* Per Olov Enquist describes the role of the playwright in the production of a play.[43] In one sense, he or she has nothing else to do once the script has been taken over by the director and the actors. And yet still he or she sits there in the auditorium during rehearsals, doing something important by the very act of being present. But what?

To answer the question, Enquist relates a trick that trotting trainers sometimes use when a horse gets the jitters and starts to gallop during training. They place a rabbit in the horse's stall. The rabbit cares little about its larger room-mate and continues to ignore it as it nibbles away at the hay. But the horse is gripped with curiosity, sniffs at its new friend and is filled with sympathy and poise. It stops galloping, calms down, regains balance in its life.

This same function, argues Enquist, is served by the playwright during production. As the premiere approaches, the tension among the ensemble grows.

The director is unhappy, the actors start to panic, conflicts erupt. The playwright is summoned. Does nothing. Just sits there in the front row looking unruffled and happy. One look at this catalyst of composure, and everyone settles down and the rehearsals can continue.

This is exactly how I see the role of the research leader. I have lost count of all the times some highly strung researcher has burst into my room announcing something along the lines of: 'I've got a huge *problem* and haven't the *slightest* idea how to sort it out!' Only to then describe the problem, explain how they were hoping to resolve it – almost always a very well-thought-out and viable solution – and leave the room, relieved and grateful. Even though I've often not done anything other than listen, hummed and nodded in assent, and merely affirmed their ideas. And in this non-way calmed them. Like a white rabbit in the race horse's horsebox.

But it is, naturally, also about being helped to see weak spots, flaws and defects in one's ideas. Mattias Marklund tells of a researcher in his own department who had found something really interesting – a potentially groundbreaking discovery, in fact. But before airing it in public he wanted to test his ideas on his fellows. So Mattias and a colleague sat down with the author one evening to go through his entire reasoning, 'supporting and criticising so that he could feel confident about having arrived at the right conclusions'.

Confidence-building criticism. That might sound odd to many people. But it is precisely this that we often find at the core of our work. When an idea is committed to paper one is responsible for it for the first time. It is a rather eerie, insecurity-engendering moment. Well-meaning and well-expressed criticism can then overcome the fear possibly building up inside us that what we are saying is trivial, or simply wrong.

And if someone takes time, goes to such pains, to not only read about what I have done but also think deeply about its veracity, and if I really have asked the right question, it can't be that meaningless, can it? And surely I can't be an imposter if people I respect can be bothered to learn about my conclusions? If they even seem to have derived something personally valuable from what I have written? Conversation and cooperation are the only weapons we have against fear and meaninglessness. In the sceptical but friendly gaze of the other we (re)appear as intellectual beings.

But such care we reserve for people we like and trust, which means nurturing the emotional relationships in the group – 'being nice is no trivial matter' as Mattias Marklund calmly states. Close cooperation is therefore only really possible among people who work well together.

These fundamental micro-processes of research, research policy has never managed to grasp.

Research Policy as Knowledge Barrier

'How do you define welfare?' The chief administrator's question is directed at me. I get the urge to tell him to shut up. Instead, I answer as well as I can, even though I know he doesn't quite understand what I'm talking about.

I feel alone and ill at ease, despite the room being almost full. In front of me is a row of blank faces: the vice chancellor's assessment group for the application for the welfare studies research centre that is soon to be submitted. None of them knows anything about my research field but some have opinions nonetheless. They have been sitting here all day long and seen a succession of the university's research luminaries bowing and scraping at the podium. Maybe they're about as enthusiastic about being here as I am. The desire to just get up and go is almost irresistible. But I stay out of some combined sense of duty and a fear of making a fool of myself.

Now the strategic planner is drawing in his pad. Here is our application, he draws. But where is the 'value added' and 'the vision'? We should reinforce that. And what does our route to our long-term goals look like really? Have we done a proper SWOT analysis? I realise he's just doing his job but I'm nonetheless filled with aggression. I don't like myself when I try to explain our thinking. Shouldn't be here. Want to get away.

The previous weeks have been spent prising out a large application for a ten-year research grant to the Swedish Research Council on behalf of Umeå University. Trying to stitch things together with colleagues who think I should work with their research agenda instead of my own. The idea of working with them for a decade fills me with nausea. But I play along in the realisation that no one else at my department will manage to clinch it. Duty trumps desire.

A few months later we get the news. No money this time. Positive evaluations, but our application falls just short of the bar. Ten minutes' disappointment and then boundless relief. I'm off the hook. Maybe it's over now?

Yes, it was over. The only permanent result of our efforts was a text on *The Bullshit Quota*. On how the ratio in an application between words of fluff ('vision', 'synergy', 'groundbreaking', 'value added', etc.) and of substance ('observation', 'analysis', 'truth', 'data', etc.) can be used as a tool for analysing research policy. The text parodies the process we had just gone through and I wrote it kind of cathartically. Perhaps as a way to recover my dignity.

None of my other texts has aroused such delight in the world of research and my inbox was soon overflowing with appreciative acclamations. A few years later it was still doing the rounds and clearly eliciting the joy of recognition. It seems to have hit a nerve. To prod a boil of research policy BS.

There is one thing on which the political left and right seem to agree, and that is that researchers need to be controlled and monitored more. This control takes different forms during different times: sometimes the idea is to herd the researchers together into 'excellence programmes', sometimes to micromanage content in 'strategic research areas'. And new ways are found of measuring and rewarding 'output'. Ideally in some automated form (e.g. citation analyses) so that no one has to go to the trouble of trying to read what they don't comprehend anyway.

A new class of research administrators has been established to take control of this machinery of steerage and evaluation. They are found in the government offices, in the research councils and, to an increasing extent, at the universities. They carry imaginative titles such as 'strategic planner', 'quality coordinator' and 'research analyst'. They are broadly speaking a parasitic group, living off the research system as they suck the energy and life out of it.

So what is happening is that energy and time is being drained out of what should be the remit of the researchers – to work together in teams to solve problems and publish results – and put into reporting back and evaluating, finding 'strategic visions' or, at best, circumventing the constant bureaucratic demand for feedback and information.

One fascinating feature of this development is that it has been so insignificantly informed by research into how research environments actually work and what makes people working in these environments work hard and achieve results. Not the slightest consideration of the kind put forward in this book features in this brave new world of research management.

Research into successful research environments identifies, as we have seen, group dynamics as the key to scientific discovery. Creating groups in which the researchers display 'Goldilocks diversity', different enough to bring each other something, similar enough to be able to communicate. And this shows how complicated it is to make such groups productive, how vital good academic management is, and how important emotional and social factors are to creating effective teams. These groups can be supported or countered, but they cannot be commanded into existence. They grow organically as the result of curiosity, truth-seeking and trust. And they are fragile. A heavy hand can easily be their ruin.

Li Bennich-Björkman writes in her book on innovative and stagnating research environments about how the growing drift of research policy towards control and evaluation risks destroying the local processes that are fundamental to creating effective environments.[44] These are built, as is discussed throughout this book, on trust and a one-sided focus on rewarding easily quantifiable indicators risks subverting an innovative and risk-taking climate.

And the constant demand for continual feedback robs what needs to be done of the time needed to do it. Research policy can easily destroy environments, but not so easily create them.

Teresa Amabile, who has spent her academic career studying the social contexts of creativity suggests that extrinsic motivation – such as interest in career, pay, good evaluation results – has a detrimental impact on creativity.[45] The researcher who cares most about the gaze of others can easily lose her imagination and powers of creation. It is instead the one driven by intrinsic motivation – curiosity, passion, the desire to do a good job for its own sake – who is creative.

Evaluations and judgements can, if done correctly and by the right people, bolster this intrinsic motivation. If done prosaically by people for whom the evaluated lack intellectual respect, it damages intrinsic motivation. When the important thing is to meet criteria rather than doing good work for its own sake, creativity is lost to the four winds.

Sometimes, researchers mount a successful resistance against control and decree. The World Wide Web emerged as an unplanned by-product at the CERN particle accelerator in Geneva. Tim Berners-Lee and his colleagues were devising a 'system to prevent the loss of information' as the prosaic description in his original memorandum reads. I've seen it myself, in a stand in the entrance exhibition at CERN. At the top of the first page his boss has written '*interesting, but vague*'. But Berners-Lee and the others soon realised that they were on the trail of something much greater and that what they had built could be used globally via the Internet to create a 'network' connecting all the world's computers in which all available information could be made immediately accessible.

However, some of their bosses were left cold by the idea. This was a particle accelerator for doing *high-end physics*, not a playpen for computer geeks. So they urged the researchers to put their time to better use. According to some accounts, it went so far that the researchers were prohibited from doing any more work on network building.

But those who were on the trail of the web refused to obey orders and continued working away on the quiet. Deep inside CERN a small secret sect emerged, eagerly occupied with building the web. They smuggled in personal computers from California in crates marked 'Apples' (which in a sense wasn't a lie). The researchers would occasionally exchange offices to prevent their bosses cottoning on to what they were up to. They refused to be governed from above and carried on drilling in what they felt was the right direction. The result of their efforts we know today as the World Wide Web, which we use every day of the year. It has revolutionised the world.

After the fait accompli, CERN has taken the creation of the web to its heart and incorporated it into the myths surrounding CERN's activities. In his memoires, Berners-Lee writes conciliatory about his relations with CERN, even if we infer between the lines that there was hardly any real support for their endeavours. But had it been up to certain research bosses, the web would have been suffocated at birth.

The head of research who tells us this anecdote while guiding us round the bowels of CERN says that he gave up trying to micromanage the researchers long ago. He says with cheerful resignation :

> Formally speaking I'm their boss, but I've realised that if I try to steer them in a direction they don't want to go in, they'll just pull the wool over my eyes. They're actually much smarter than me. So I reason that if I let them carry on with what they find most interesting for now, I can be fairly sure that they'll come up with interesting findings.[46]

Research policy seems driven all too often by a fear that researchers will lose 'relevance' if they are left too much to their own devices. This relevance, as far as social science is concerned, is often manifested in the demand that the research must result in marketable products, solve pressing organisational problems or relate to issues of current affairs. Maybe this research policy vacillation comes from the fact that those who take the policy decisions often lack personal experiences of well-functioning research environments. Either they have never been researchers or their experiences have made them regard the university as a stodgy, dusty ingrained place in need of a good shake-up with clear criteria and quantifiable goals.

There is a worrying lack of interest in thinking about or examining how the microprocesses of research environments are affected by policy interventions. Are large research centres to be preferred, or many small projects? Is it better to provide long-term support for the few, or short-term financing for the many? Are open calls for application better than thematic, intradisciplinary programmes better than interdisciplinary? Better in what way? Is there a danger that policy clumsiness will simply damage well-functioning research environments? No one knows. And what is worse, no one seems to want to know.

But this is not the worst of it. It's impossible to find even the slightest indication that anyone is thinking, or at least trying to think about how a knowledge-producing environment actually works. What incites researchers to make an effort, to think creatively, to help each other, to break through barriers? Is it really the hunt for higher 'mean citation'? What does a well-functioning research team look like? What relationships characterise it? How does policy impact on the researchers' joy, passion, cooperation, competition? Not a word. I asked all researchers I interviewed or talked to for this book if they could recall a single passage in a policy document that touches on such issues. Nothing. No one can remember ever seeing anything of the kind. They smile a little awkwardly and agree that it is actually rather strange. They know themselves that this is where the key is.

Sometimes it descends to parody. My one-time EUI colleague Chris Reus-Smit tells me with a mix of delight and alarm about the latest evaluation at his home university in Australia – which excluded the real star of the department. In the mid-1980s, this person wrote a pioneering work that everyone in the same field is required to read and cite. Then he published almost nothing for two decades until a year or so before the evaluation, when he came out with another possibly seminal book that blew open a completely new research field. Yet he still could not be included in his departmental evaluation since it confined itself to those who had published at least three peer-reviewed works in the past two years. The fact that he'd written two books that had transformed his entire research field did not come into it. They wanted three, and recent to boot.

When assessors of research are handed such directives, it's perhaps not so strange their work is usually greeted with contempt by the researchers. As something to be ignored, circumvented, outwitted as much as possible. *Those who can't do, evaluate.*

Research leadership and financing should, instead, operate roughly like the terrorist network al-Qaida – if you allow me such a perverse analogy.[47] They should form a base to which people with ideas go to obtain resources and feedback, not someone high up who tries to govern them by decree. As the world discovered in the case of al-Qaida, this kind of seemingly leaderless organisation can be incredibly effective. 'Organisations set up to solve complex problems must be loosely coupled' as organisational research puts it. Research policy makers, in this regard, have much to learn from Osama bin Laden.

Ultimately, it is a matter of safeguarding research practice and the intellectual craft as an 'endangered species of practice', as the Danish planning researcher Bent Flyvbjerg so neatly puts it.[48] Before the evaluation frenzy and the tyranny of the immediately measurable, one must assert the primacy of lived, embodied research practice. The context in which knowledge is produced by living, fragile, complex human beings.

Because Hubert Dreyfus is right when he says in interview with Flyvbjerg that 'disciplined bodies will not be as good at what they do as intuitive bodies'.

And the researcher for whom 'mean citations' and good evaluation results have become the sole focus of concern will not be exciting for much longer. When I hear the word 'output' in the same sentence as 'research' I just want to cock my revolver.

German sociologist Jürgen Habermas captures the dilemma of modernity in his 'system and lifeworld' concept.[49] State and market have been severed from the everyday contexts in which people are embedded and in which we create relationships and meanings. This has achieved greater effectiveness in the state's calculating decision-making and the market's impersonal utility-maximising mechanisms. But state and market – the system – constantly threaten to colonise the lifeworld and undermine, destroy even, the vital processes that go on within. Interhuman relations become commodities, priced, or the grounds for the exercise of power and manipulation. Ultimately the system is also threatened, for without the basis of a functional lifeworld it is unable to procreate.

Similarly, the evaluation machinery now set in motion, and the research-governing motives it embodies also threaten the everyday practices without which new knowledge cannot be generated and preserved. The interhuman contexts based on trust and sympathy, which are the prerequisites of every functional knowledge-producing context, are undermined by faultily designed research policy measures. The lifeworld of research is under threat, and this should be a cause for scholarly concern.

Should researchers not be more vociferous in their protests against this development? Sure. But we are opportunists. We want to research, and research costs money. If money is made available, we try to obtain it. If possible, in the form we believe in. If not, in the form on offer.

So should research policy not be better suited to the actual conditions that prevail on the front line of knowledge production? This would require a better understanding of the conditions that prevail where knowledge is to be produced and complex problems solved, better insight into how important human relationships and emotions are in these contexts, a less *von oben* view of the research landscape and those who roam it. A better research policy requires therefore a position of trust towards researchers. Faith in our ability to prioritise and deliver. Prudent support for the environments that have established themselves organically and proved successful. Investment in large (but not too large) programmes for large (but not too large) groups. And no more meaningless evaluations.

But would this not lead to researchers becoming lazy and indifferent? Here one must be fully aware that internal scientific criteria are based on control, criticism and competition. Incentive is built into our own career and status systems. We compete from the day we are employed until the day we retire: for positions, publication space, scant resources. We can prioritise, we see what is relevant, we are dynamic and inspired. When we are our best. And if we could just be left in peace to do our job.

Nothing seems so inadequately research-governed as research policy.

Chapter 2

IN THE CORNER

This is your last chance. After this, there is no turning back. You take the blue pill – the story ends, you wake up in your bed and believe whatever you want to believe. You take the red pill – you stay in Wonderland and I show you how deep the rabbit-hole goes.[1]

Inside but Outside

It looked so strange. They'd climbed up the big tree, just outside the gym. They were now looking in on us as we busied ourselves about the room. The sun had started to set so they were barely visible. But if you looked closely enough you could see that there were many of them, staring curiously, and everything was pretty strange. My little brother and I continued to try getting the basketball in the net. But we were feeling uncomfortable now, with our unwelcome spectators. My mum was doing sit-ups, what my dad was doing I don't recall. But it was thanks to him that we could be here, on a Saturday, even though the school was locked up. He had a bunch of skeleton keys, since he was a teacher. My mum was also a supply teacher there, but I don't think she had the keys to the gym. My dad was an important person in the village: his opinions counted and it was to him that people turned if the villagers' views had to be committed to paper. I realised that was why those kids sat in the tree watching us: because we were special. Because we belonged and yet didn't. Because we were outside, but on the inside.

Many years later when I read sociologist Dalton Conley's autobiography *Honky*, in some strange way I could relate to it.[2] And yet the childhood he describes, as the only white kid in a class in a violence-prone New York slum, is possibly as far away from mine as you can get. Back then, the little village of Jockfall just north of the Arctic Circle boasted roughly two hundred inhabitants – these days I daresay there are no more than fifty. But the way Conley describes the position he was in, with his white skin and artist parents, as being part of the urban slum and yet not, called to mind the picture I have of my own young life in Jockfall. As the son of teachers in a village of foresters. Observed and treated differently. Marginal but not outside. Vaunted yet at the same time slightly despised. This, too, a future sociologist's upbringing.

I have returned to this position time and time again, as if caught in a repetition compulsion. It took me years to see the pattern. That I was constantly distancing myself from the situations I was in; that I preferred to place myself on the periphery, with a full view of proceedings but without really, fully participating in them; that I always managed to find a position a little to one side, a little elevated, special. From the banal – about where my office was, preferably on the fringes of my department – to the altogether life-defining, in relation to friends and social life. A certain solitude has come from all this: I have many acquaintances but almost no really close friends. But I can't be anyone other than who I became, and the privileges that have fallen to me along the way have been sufficient in number to compensate for the loneliness.

In this part of the book, I will be discussing *marginality* and its significance for the scientific intuition and creativity, in various ways and from different vantage points. Many – but not all – of my examples come from social science research. This is mainly because that is where I have gathered my own experiences and observations. I believe that much of what I have to say is just as well suited to natural science research or to other innovative occupations, such as writing, painting or carpentry. At the same time, the character of social science research is, I think, a little special in that the studier is, to some extent, also the studied. That the produced also includes the producer. This makes, as I shall often reiterate, the importance of creating and maintaining marginality particularly salient here.

My contention is that some kind of marginality is prerequisite to the ability to analyse social contexts and processes in depth – indeed, even to be creative in whatever occupation one might choose. Some people have been handed this marginality, free and unasked for, along with their upbringing. Others have created a marginal position for themselves through their careers or the way in which they organise their work. I will be giving examples, describing how marginality can be deliberately created and upheld, calculating the price it comes with.

Let me stress here and now that what I mean by marginality is not the same as being on the outside or excluded. Whoever is placed, or places themselves outside – who is excluded, to use a social studies buzzword – has left or never been allowed to enter a situation. They have been expelled or barred from or exited of their own accord the room in which the intercourse is playing out. However, the one who resides in the margin has not left the room, only ended up, willingly or not, in a corner, away from the hurly-burly of the centre.

If this person is so inclined, he or she might scan the room and discover that what is going on is so much easier to see from the corner, that what was recently a confusion of events, people and bodies can now be seen as patterns and recurring chains of events. That what is happening is even explicable. Whoever wants to discover something genuinely new would therefore be wise to stand in the corner, and neither let themselves be hustled out nor drawn into the happy maelstrom in the middle. But the people in there seem to be having so much fun, so if you want to stay in the corner to see more clearly, you will miss out on a good deal of fellowship and joy. I have found that to be a price worth paying, but I'm not sure if my calculations are correct.

A distinction can be drawn between marginality as an innate social position and marginality as a designed product. In the former case, it is about how certain people – by virtue of their ethnicity, religion, class or whatnot – have come to occupy a special kind of marginal position in the social rooms in which they find themselves. Like that which my family and I had up in Jockfall. Slightly more dramatically, we can think of it as 'the generalised Jewish experience', as a reflection on the singular social status that the Jewish population had in Europe before the Nazi genocide.

Much has been written about this kind of marginality in my own academic discipline of sociology.[3] Maybe it is even the case that sociology's position as a kind of fringe or mixed discipline with diffuse contours and a nebulous core makes sociologists particularly fascinated by other marginal phenomena. The seminal text is an article from 1928, in which Chicago-based sociologist Robert Parks coins the term 'marginal man'. The, let's now say, 'marginal person' is characterised as having one foot in the old, time-honoured and familiar, and one foot in the new, negotiated, unfamiliar. We are talking twentieth-century migrants from the American South to Chicago's bustling city life; the women who took the half-step from a life centred on home and children to a career; members of ethnic groups on the fringes of mainstream society; and many others. According to Parks, the marginal person is characterised by an insecure social position, by no longer being part of their beginnings but not fully part of their destination either.

Even if Parks and his many followers stress that this social position can be the source of creativity and innovation, it is nonetheless the negative aspects of the marginal existence they emphasise most strongly. The marginal person is tormented by an existential insecurity that leaves them susceptible to mental, and ultimately social problems: rootlessness, disorganisation, destructive conflict. This was framed in rather dramatic terms by a sociologist, who in the early 1950s expressed concern at the situation of the 'marginal woman', she who had taken the step from a life of pure housewifery to one in which home and work had to be coordinated: 'Uncertain of the ground on which she stands, subjected to conflicting cultural expectations, the marginal woman suffers the psychological ravages of instability, conflict, self-hate, anxiety, and resentment.'[4] The stifling tedium of suburban life from which this marginal woman had just liberated herself, however, is left uncommented. Her previous confinement and the circumscribed living spaces that films like *Revolutionary Road* or novels like Marilyn French's *The Women's Room* depict so powerfully is given no place in the discourse on the predicament of the marginal woman.

I do not wish to play down the psychological alienation and conflicted role expectations that can encumber a social position in the margins – or the difficulties of piecing together what has become known as the 'life puzzle'. What I am proposing is that the costs of marginality are greatly exaggerated in the Parks tradition and that these costs must not be allowed to obscure the positive sides of marginality. The opportunity to stand on the sidelines opens up new social spaces, allows us to gather resources and perspectives from disparate worlds, to piece things together in new ways. After all, there were few women who genuinely longed to return to life as a housewife, even when their new role expectations seemed impossible to fulfil. Is it really peace we want? wondered the Hoola Bandoola Band (and later Imperiet) once. No, maybe not, and even though a life on the margin is in many ways a more strenuous life than the serenely embedded, it can also offer dynamism, excitement and unexpected shifts of perspective. To stand on the edge is to not leave yourself in peace, to never settle down, to forever seek.

The literature on the marginal person focuses mainly on the marginality one is born into, or unintentionally placed in by movements in geographical and social space. But marginality need not be one's destiny, but something one achieves and sustains, a designed product rather than an imposed existence. In this case it is – wholly or partially consciously – about creating and maintaining a marginal position relative to knowledge field or social context. Here, rather paradoxically, marginality is dependent on the social and emotional relations we have to the other actors in our environment, on how we can distance ourselves from our friends and take care of our foes. For in the words of Friedrich Nietzsche's fictional prophet Zarathustra: 'The man of knowledge must be able not only to love his enemies but also to hate his friends.'[5]

I hold that marginality is a condition of genuine creativity. But what exactly *is* creativity? This word that is abused so often that it is now almost bereft of meaning. What I mean by creativity is quite simply the ability to create something genuinely new. Naturally, whatever this 'new' is, it does not have to be a physical object. It could equally be a piece of music unlike anything ever heard, a thought or a perspective made accessible or expressed for the first time, or the framing of an idea that makes us see it in a new light. But it can also be a way of working and assembling pieces of wood into a chair that we have neither seen nor sat upon before. And while the act of creation rarely starts ex nihilo – it is often a matter of combining pre-existing elements in a novel way or improving pre-existing processes – the creator adds something that we have never previously seen. Creativity has an elusive character, existing in the eye of the beholder, existing only if accepted by others as creative and useful to their own endeavours. Much could be written about this and indeed has been, but it is not really what this will be about. Creativity is the ability to create something new; no more need be said.

Born on the Margins

Our family never spent any summers in Jockfall. On the first day of the summer holidays – which my teacher parents could also enjoy – we headed for the summer cottage only to return a day or so before the start of the autumn term. To be sure, the cottage was only 50 kilometres away, but it was still remote enough for us to never make more than the odd trip home to cut the grass and check the place over. The holiday adventures the other children related to each other when we were back at school were alien to my brother and I (and my sister, who was too young then to have any memories of Jockfall). Our adventures were different and seldom told. I remember the occasional visits to the summer-warm and fairly desolate Jockfall, how strange the empty school yard felt, how I both wanted to stay and to desperately get away. And how the other children always seemed a little strange at the start of the school year. Like a different species almost, yet not.

In Jockfall, most people were related to each other in one way or another. They were mainly foresters and their families, or (less commonly) small-scale farmers. This did not mean that the village lacked conflict or was even particularly harmonic. The native urbanites who sometimes get it into their heads that life 'in the country' is an idyll have presumably never tried living there. A depopulated area like the interior of Norrbotten is, moreover, the victim of a rather pronounced form of negative selection. Many of those who had the will and skills moved south; the ones who remained were sometimes of a rather problematic kind. True, the vast majority of the villagers were peaceable, hard-working people, but there were a striking number of individuals who had gone off the rails in one way or another. Even though it was really only a few families who were this way, they made a profound mark on the way I perceived the village from my childhood horizon.

My worst childhood tormentor died in his early twenties after trying to booby-trap his love-rival's car, only to blow himself up in the process. Either that or he took his own life by this dramatic means – the stories differ. My childhood was full of unpleasant but titillating stories of drunken and armed sibling showdowns, thefts and burglaries, child abuse. Everyone knew which kids were being beaten – it was almost always some alcoholic father who meted out the abuse – but no one intervened. I shared a class and a school with many of these unfortunate children. And you had to watch out for them: they tended to take out their distress on their friends.

In this environment, the teachers and their families were a rather odd feature, an imported element of an otherwise fairly homogeneous village. Dropped in, as it were, by parachute, far from the front line, with a mission to bring learning and enlightenment. Both my parents came from meagre, but not impoverished circumstances. My grandfather on my father's side was caretaker at Överkalix court-house, his wife a seamstress and cleaner; my grandparents on my mother's side were smallholders. In many respects, the social gap between by parents and the rest of the villagers ought to have been more of a crack. But the mere fact that they had gone to secondary school, that my father had even attended teacher training college, was an unbridgeable crevice. My father could express himself in speech and writing and helped to found a new political party; my mother had shelf after shelf of literature at home. It wasn't like that in the other families.

We were fairly well off compared with the others in the village, teachers' salaries being considerably more than anything piece-working forest workers could scrape together. Our home was popular with the other children; not for its inhabitants, mind, but for the many comics and toys it contained. I would commonly leave the other children there and simply walk away as wandering around the forest or the village was far more enjoyable to me than their company. My abandoned friends didn't miss me that much, content as they were in the company of the comics and toys.

As an adult, I've often wondered how this childhood environment shaped me. That I even had a distinctive environment only struck me when I was 14 and we moved to my father's parental home in Överkalix. My grandfather had died of a heart attack the year before, and as my grandmother couldn't take care of the house on her own, the natural thing was for us to move in. Only a few dozen kilometres separated a village of 200 residents from a small town of over 2,000. But what a difference! Here, it wasn't that remarkable to be the child of teachers, and I was one of many. People had all sorts of jobs, even though not a single person could have been said to be upper class. After a short time as the new kid on the block, no one thought I was odd. I played basketball, rode a moped, went skiing. Just like the others, one of the others.

And yet not. I soon discovered that here too I had some kind of special status, but one based on my being a new arrival – and thus not really part of the cliques that the other kids were in – rather than on being the 'teacher's kid'. It wasn't as if I was excluded, quite the opposite. I was the only one who was accepted in both the somewhat hostile camps that most of the boys in my class had grouped themselves into, so much so that once or twice I served as peace broker when their rivalry grew too heated. I was the only occupant of the subset where the two groups overlapped. And also a little odd, with my overly eclectic interests. Mum's bookshelves had left their mark.

And thus has it gone on, in circumstance after circumstance, from boyhood to adolescence to adulthood to middle age. With a few brief exceptions (a couple of years as a university student) I've never really been fully able to integrate. 'So how *is* it with you and collectives?' as a left-wing friend asked with somewhat friendly irony when I once held forth about the horrors of camping holidays. Supporters' clubs, political parties, parents' groups, neighbourhood associations: they have all filled me with tedium or disgust despite knowing how much they have meant for others and for societal organisation. It is, sorry to say, only when I myself have to lead a project that I can settle down into the collective, despite genuinely needing the others for my thinking and well-being. And I'm a good leader; whoever once brokered peace in Class 9A has it in their bones. But to happily be included as one of many is beyond me. Jockfall lives in me still and renders it impossible.

The kind of marginality that my upbringing created can be found in many other places, with its roots in many other kinds of experience, often of a much more problematic variety than mine. Sociologist Dalton Conley's autobiographical discourse, which I have already mentioned, tells of a childhood at once completely different to mine and yet so structurally similar. Dalton was the child of culturally practising parents, who had concluded that the only way to devote themselves full-time to painting and writing was to move to where the rents were lowest, which is to say the tower blocks of New York's 'Projects'. Here the young Dalton stuck out as the child of white middle-class parents.

Back then, the young Dalton still found it hard to grasp that skin colour could mean anything other than just that. He tells a poignant story of how he longed for a little sister, how his mother made it clear to him that a sibling would somehow arrive in the family; but when would she come, the little sister he so pined for? Then one day he caught sight of her in a pram in the playground, a little black girl smiling up at him. There she is, he realised, I've found her, in a pram in the playground. He grabbed the pram, rolled it away across the playground to his mother, to whom he merrily presented his new-found little sister. The horrified mother then had to explain the whole thing to the little girl's angry parents. The improbability of Dalton's chalk-white parents having a dark-skinned baby never occurred to him. His social world was organised by size, not by skin colour. And his prospective little sister was, just like him, little. Thus they belonged together.

It isn't long, however, before Dalton understands that skin colour often means much more than mere differences in pigmentation. The black children in his class are, for example, slapped by the teachers whenever they misbehave, for it's what their parents want in order to have their offspring licked into shape. But the school realises that Dalton's liberal parents refuse to accept corporal punishment, so he becomes the only child in the class whom the teachers aren't allowed to touch. And so on it goes, in one walk of life after another. The white boy is there, but on the periphery, not really one with the others yet part of the context nonetheless. It slowly dawns on him that skin colour organises the world and that he himself occupies the fringes of his own environment.

Sometimes, marginalisation proves a blessing. Young Dalton is drawn to a gang of violent delinquents, is rejected at first but is then allowed in because they want to use his new baseball glove. He's one of the gang for one single day, until they simply take his glove and kick him out. Dalton protests but a knifepoint to the throat changes his mind. A horrible day, a horrific experience. But as an adult he looks back and wonders what became of the others in the gang: prison terms, drug abuse, early death. What would have happened to his younger self if they hadn't kicked him out? Would he also have jumped onto the slippery downward spiral that became their life? Dalton Conley considers himself lucky that he wasn't allowed in.

Gradually, it also dawns on the Dalton's slightly naive parents that the school environment they have put him in is directly harmful. Via their contacts, they manage to bogusly register him as a resident of Greenwich Village, where he enters a middle-class school. Dalton now discovers 'class' for the first time. In the Projects no one ever had any money, so no class differences could be discerned. But now suddenly Dalton is one white kid among all the other white kids, but the only one who never has any money and the only one who returns, when the bell goes, to the poverty and crim- inality of the Projects. He's back in again, but not really. And in the back- and-forthing between the colour-based marginality in the Projects and the class-based marginality of Greenwich Village, the seeds are no doubt sown of what will one day be a luminous career in sociology. In the gaze from the sidelines, from the corner of the room. To precociously learn that nothing can be taken for granted. 'If the exception proves the rule, I'm that excep- tion', he writes laconically and perhaps a little sadly in summing up his childhood.

The marginality we experience as we grow up can take different forms. Sometimes it is to be found within the family. In a conversation with fellow author Stephen Fry, Malcolm Gladwell talks about his parents and what his family background meant for the way he thinks and writes.[6] His mother is social worker and psychologist from Jamaica, his father a British mathematician. Along both dimensions – ethnic and professional identity – distinct variation, something that must constantly be lived and negotiated within the parameters of the family. What was valuable, what experiences were important, how was life to be lived?

What has this meant for you, wonders Fry. That I have learnt to look for uncommon commonalities, replies Gladwell after a beat's reflection. Being able to see links between apparently altogether different or incompatible elements is naturally one of the fundamental components of the creative approach.

That society cannot be taken for granted is probably the first lesson one must learn before starting to seriously understand how it actually works. And that insight perhaps comes most easily to the person whose own experiences tell them that society can actually break down and devolve into chaos, cruelty, violence. Author Göran Rosenberg says in an interview how his family history – his parents were Auschwitz survivors and his father later committed suicide – influenced his own approach to life:

> By imbuing me with a strong consciousness of society's fragility. I am constantly aware that human society is always in danger of disappearing or falling apart or becoming cruel, and that a serious effort must be made to maintain a tolerant, open society. Many Swedes were born with a sense of security – this year Sweden celebrates 200 years without wars. But I feel different. I do not feel secure, I feel that everything is fragile, temporary, poised on the brink.[7]

Being poised on the brink was for long the Jewish experience in Europe. Historian Eric Hobsbawm reflects upon the over-representation of Jews among Nobel laureates, lauded composers, pioneering authors, outstanding musicians and others with pronounced creative professional successes.[8] According to Hobsbawm's essay – as well as in the terminology of this book – it is when exclusion is replaced by marginality that the fantastic Jewish successes emerge; when the Jews in Europe start to use the national language, are allowed into higher education and move beyond the ghetto and a narrowly defined field of business, while never fully being at the epicentre of politics and culture. Exactly the kind of marginality that has always aroused suspicion or even hatred in those who seek uniformity, purity, clear boundaries. As its most explicit manifestation, this hatred breeds a desire to eradicate the Other from our midst.

But at the same time, this is the same marginality that increases the chances of viewing things in a way that no one else does, of finding new combinations and new avenues to development. 'It is as though the lid had been removed from a pressure cooker of talents,' Hobsbawm writes, encapsulating the veritable explosion of Jewish scientific and cultural achievements that followed their emancipation.

He also points out that, interestingly, there is no over-representation at all of Israelis among the Nobel laureates or other paragons of the scientific-cultural world. This throws a spanner in the works of suspect genetic arguments about Jewish successes (such as those propounded in the best-selling but highly dubious *The Bell Curve*).[9] It also puts into perspective one-sided religio-cultural explanations, which ascribe their scientific and cultural accomplishments to some aspect, unclear what, of the Jewish religious tradition. But this should then operate just as much on Israeli soil, should it not?

For Hobsbawm, just as it is for me, it is instead a matter of marginality, of constantly navigating the outskirts of a society that others can always take for granted, finding creative compromises that without erasing one's own individuality allow one to live with the others. And of always being made to see that neither one's own nor other people's way of life is natural and unproblematic. It is this, having to live as a minority among non-Jews, that creates this singular position from which pioneers are born, this 'lack of fixity', this openness to innovation, writes Hobsbawm.

A similar experience is captured in autobiographical form by another great contemporary historian – Tony Judt – in an essay succinctly titled 'Edge People'.[10] In it Judt reflects on his own childhood as a Jew in early post-war England, on how 'the warm bath of identity' was never his, how he always thought of and wrote about the English as 'them', rather than 'us'. But also how the Jewish experience and identity, far from being deep, rather left him 'a decidedly non-Jewish Jew'. Neither Englishman nor Jew yet both – an edge person. In adulthood, this upbringing expresses itself as a distinct penchant for the margins and niches:

> I prefer the edge: the place where countries, communities, allegiances, affinities, and roots bump uncomfortably up against one another— where cosmopolitanism is not so much an identity as the normal condition of life.

In, but out. And despite the ocean of experiences separating us, the Jockfaller in me nods in recognition.

I meet my good friend political scientist Bo Rothstein to chat about his Jewish roots and how they shaped his life as a scholar. I'd say that Bo is the most creative person I've ever had the pleasure to work with, and believe me, I've worked with a good many creative people over the years. A never-ceasing torrent of thoughts, ideas and arguments flows constantly from his head. They are discharged, examined and repeated in a way that can be quite trying for others but is nonetheless so stimulating that it feels like a privilege to be in his company. Not all his ideas and arguments are brilliant, but frighteningly many are. And the publications, citations and prizes speak for themselves: Bo Rothstein is an altogether extraordinarily successful scholar. But where does it all come from?

Bo describes his upbringing as one with 'one foot in two different camps' in a rather strict Jewish family in an otherwise religiously indifferent environment in boom-years Malmö. Living in accordance with the orthodox Jewish tradition with its many edicts and prohibitions had made a modern life virtually impossible. At the same time, Bo's family were reluctant to relinquish their Jewish allegiance, to renounce the core values of the tradition they had grown up with. The consequence was constant creative compromise. Walking the 6 kilometres to the synagogue on a Saturday, which doctrine decrees since driving is not allowed on the Sabbath, would be a rather impractical and distinctly arduous experience in the sleet of a Swedish January. But driving the whole way would be both morally wrong and unnecessarily provocative. Their workaround was to drive the first 5 kilometres, park and walk the rest. Hypocrisy? Well, maybe more a way of compromising between two irreconcilable worlds.

Bo laughs about how his mother developed an unfortunate taste for smoked eel, an animal that a devout Jew is forbidden to eat. How was she to indulge this passion? By making a decision that the eel was not to be eaten on porcelain in the kitchen or dining room, but only on a paper plate out in the garage. The image of Bo's mother, eagerly tucking into her beloved smoked eel from a paper plate in the family garage is a kind of emblem of marginality's creativity.

'We weren't at all marginalised in the sense of being outside,' says Bo, whose family always saw themselves as wholly Swedish, only with a religion that meant they had slightly different habits and traditions to others. They lived as it were on the edge – 'on the threshold' – but felt in no way excluded or in the slightest bit discriminated. This was long before the Israeli–Palestinian conflict became an inevitable touchstone of the Jewish experience, and long before identity politics with its focus on origin and affiliation elbowed its way into the debate. There was no talk of the mass-killing of Europe's Jews, even though his paternal grandparents were murdered in Chelmno extermination camp. Maybe the crimes and experiences were too terrible to even be mentioned. It was not until his teens that Bo woke up to what had actually happened to his grandparents and six million other Jews.

How did all this influence Bo's intellectual achievements? He gropes around in the conversation, stressing that an upbringing such as his is no guarantee of creativity; others who share it have developed no particularly creative take on life at all. At the same time, the aggregate numbers speak for themselves: a distinct Jewish over-representation among the pioneers of the sciences and the arts.

But in his own case? He doesn't rightly know. 'Anyway, I decided early on that I'd never research anything to do with the Holocaust. I wouldn't be able to keep any kind of critical analytical distance, it was simply too emotionally close.' But maybe his interest in studying the machinery of the state was nonetheless piqued by reading Raul Hilberg's *The Destruction of the European Jews*. Here was born an insight that to understand politics, one should 'roll it up from the back' – in other words, focus on what the state actually does on its outflow side rather than exclusively on elections and parties and ideologies. It is here, in the capillaries of state, that the core of politics becomes visible, be it the Nazi genocide machine or the Swedish reform bureaucracies that became Bo's first research project.

But I daresay it's not primarily in his choice of research object that Bo's background shaped his intellectual achievement, but something altogether different. To get there, our conversation apparently takes a detour.

'I'm the kid brother of the '68 generation, aren't I?' laughs Bo. 'And this had a much stronger impact on my intellect than my Jewish background.' This was on a purely personal level, his big sister's radically Marxist boyfriend having become an admired referent. But it was also the case in the professional realm, where the sociologist and Marxist figurehead Göran Therborn employed the young Bo on a large research programme on the history of the Swedish societal model.

The problem was that young Rothstein had not taken a single class in sociology, so a doctoral position in the subject was out of the question. Instead, it had to be political science, which, after all, was his degree subject. However, at the department of political science in Lund, Marxism was a dirty word, and there wasn't a single person who could supervise him through the thesis that he had now begun under the aegis of the project. He therefore had to apply to a third department, this one in Copenhagen, where Lennart Lundqvist, all but banished from Lund, became his supervisor.

Financed from one quarter but with formal departmental affiliation in another, and his supervision in yet a third, Bo embarked upon his academic career. And these were three environments that had very different views on society, 'like night and day'. The theoretical perspective that was considered a given departure point in one was dismissed by the other as sheer ideology production. What was scientifically de rigueur at one department was spurned as uninspired number-massaging by another. What was the core of politics here, was not there. 'I could have been ground down,' says Bo, but for some reason he was spared this fate and was able to exploit the crossfire of perspectives and arguments he had ended up in, and learnt to constantly see how 'the other' thinks, to constantly be confronted by radically opposed worldviews and perspectives.

I would say that it was in this regard that Bo's special upbringing had its significance. It was of course easier to ride out this slightly chaotic situation for someone brought up in a world of continual and clever compromise, for someone accustomed to routinely accommodating contradictory demands rather than clear-cut allegiances and taken-for-granted rules. For one such as this, it is probably generally easier to also accommodate the intellectual complexity of having to listen to mutually irreconcilable viewpoints, to occupy this overlap with resolve and fortitude, in the margin that has so often been the Jewish population's place of residence.

But marginality can, of course, derive from circumstances other than religious background. An interesting boundary experience comes from the family of neuropsychiatric conditions that are given such labels as 'Asperger's syndrome' and 'high-functioning autism'. Such people are often highly intelligent but have brains that process information differently to us 'normal' (or 'neurotypical', to use the more neutral term) folk. This commonly causes difficulties in handling sensory impressions, reading social situations, understanding subtexts and irony, or interpreting body language. In a nutshell, a cluster of problems that tend to hamstring social intercourse, particularly during the awkward phases of childhood and adolescence.

At the same time, many of these people have an obduracy that rarely allows their opinions and views be determined by what others say and think. It is precisely this kind of pertinacity that often lies behind intellectually groundbreaking discoveries and insights.

One might say, to me a little crassly, that the price of being in the margin is not as high for someone on the spectrum. They are not as dependent on the approval or even acceptance of others in order to feel good, as so many of us others are. For them, it is easier to live by Nietzsche's maxim to 'love their enemies but also to hate their friends', even though it is rarely a matter of hate and love in any direct emotional sense. The social exclusion that is a sword of Damocles for those who always speak their mind – even when it is uncomfortable to do so – does not pose such a threat to people who could easily live with themselves and their thoughts should the need arise. The anxious validation-seeking that is ultimately rooted in a fear of abandonment can more readily be avoided by those who are emotionally autonomous.

In her fascinating autobiography *Real Person*, author Gunilla Gerland describes what it can be and feel like to dwell on in the margin in more sense than one. She was brought up in a chaotic family in which alcohol, drugs and physical abuse escalated to the point of grotesqueness. School was baffling and pointless, friendships either non-existent or destructive. Gunilla's later social problems were related naturally enough to these arduous circumstances, and many were the failed attempts to understand and remedy what was assumed to be childhood trauma with conventional psychological theories and methods. It was only much later that it became obvious that something else was at play, and at around the age of 30 she was diagnosed with 'high-functioning autism'. Gunilla's experiences include everything from being a junkie on the streets of Barcelona to becoming a lauded spokesperson for people with neuropsychiatric disabilities. (A young woman with Asperger's once said that 'Jesus and Gunilla Gerland' were her greatest role models).

The absurd existence that was Gunilla's childhood is described in her book with an objectivity and dry humour that make it a reading experience out of the ordinary.

She describes people as 'very naive in the way they simply took over the values of their parents' and 'rather pathetic in their need to be loved by everyone' and almost feels a little sorry for them for not being 'able to disregard their own feelings, to keep things and people apart, even' when they try to think and reason.[11]

For many years, Gunilla was driven by a desire to fit in, to be like everyone else: 'I didn't miss other people, but I missed not missing other people.'[12] Carefully observing what others did and thought was initially a 'way of trying to teach myself to be like them', but eventually it became a way 'to understand what I have in common with others and what distinguishes me from them'.[13] She is now no longer that interested in being like other people, more in trying to understand how they thought and why they did not think like she did. Like a social anthropologist observing a lost tribe.

I first encountered Gunilla Gerland in the mid-nineties, when the final draft of *A Real Person* was ready but her own future seemed unwritten and problematic. Over the years, I've followed her career from a distance, the journey from marginalised nobody to a deeply respected professional in the field of autism-related disabilities. In the autism movement, people with their own diagnosis have demanded just and fair treatment, respect and adequate support, and Gunilla soon emerged as perhaps the most eloquent spokesperson of all with her linguistic brilliance and phenomenal ability to always keep things and people apart. From here, her trajectory led her to her own career in the field, with education and authorship on the programme.

In the hot summer of 2014, I call her for an interview, being keen to know more about what she has to say about her life on the margin. Since we last met, her life in the field of autism has taken a rather peculiar turn. To explain: in recent years researchers have discovered a disorder, Paediatric Acute-Onset Neuropsychiatric Syndrome (PANS), that can be triggered by an infection and cause the immune system to attack, so to speak, the brain. If a child contracts this disease and remains untreated, the chronic symptoms are very much like those of autism, which can tempt a misdiagnosis. Gunilla has not only done much to raise awareness of this fact, she is now convinced that she is actually one of the misdiagnosed herself. We chuckle at the fact that such a high-profile polemicist as she has been probably never had the disability that everyone, including herself, assumed she had.

But a life on the margin it has been, regardless of the reason. 'I've never been a run-of-the-mill type,' declares Gunilla, who has always been her own person in whatever situation she finds herself. Apart from as a professional authority on and debater of autism-related issues, the public has also seen Gunilla in the role of spokesperson for the right of single parents to have assisted reproduction treatment. And in a diagnostic examination, she was found to be 'uncommonly gifted' and among the tiny proportion of the population at the far end of the IQ curve. I'm happy to accept the test results since I've rarely conversed with someone who radiates such lightning-speed intelligence. As a girl, she tells me, she was surprisingly brilliant at certain things while totally useless at others. Life as a 'prodigy' is also a type of marginality that Gunilla has very much been aware of throughout her life.

A life on the margin in more sense than one, in other words. But what is it that people on the margin actually do? Gunilla wonders. To her, the main thing is about being loyal to one's own thinking, not other people's. It is not about lacking emotional ties to other people, but about not being as eager, like so many others, for people in general to accept what one does or thinks. It is precisely this that make it harder, if not impossible, for many to stand by their opinions even if they turn out to be unpopular. It also means that a person on the margin often 'can't help themselves' and lets their thoughts and their arguments run their course and land in their inevitable conclusion, no matter how awkward or disquieting it might be.

Gunilla is amused at the various reactions she has had over the years from fellow autism-professionals when someone like her 'from the patient side' transpires to be the best informed and most articulate. Some think it is unforgivable to have their world view 'spoiled', while others are enchanted. No matter what, it will not keep Gunilla quiet: 'I don't want *everyone* to like me, I'd not feel healthy if everyone liked me'. For many others, loyalty towards the 'icon' 'blinds them to what's bad, to what's wrong'. And it's sad to think that there are 'so many people who have so much to say that never gets said, simply because we have such a dominant consensus'.

Perhaps what Gunilla has to say about the meaning of marginality is all about the importance of being able to *be* awkward without *feeling* awkward; about standing up for what you believe is right even when the whole world says you are wrong; about not bowing to consensus. The margin is where the voice of truth is heard, even if the person on the margin is not always truthful.

Even though I myself have been a professor for over two decades, and even though in many people's eyes I am as close to the beating heart of the academic world as anyone can get, I can still, even today, find it all a little alien. What are they doing, really, these creatures of planet Academia? How can they take everything so deadly seriously? Are their deeds and desires comprehensible? When discussing the university I once worked for, I've almost invariably referred to its denizens as 'them', not 'us'. I know that this is a feeling I share with many other ostensible insiders, but the fact we share our outsider-ness makes it no less real.

For a long time I considered this a shortcoming, something I ought to conquer so that I could eventually settle into the academic world. After all, I knew that research was what I would devote my life to given half the chance. And it was at university that this chance existed, so could I not genuinely try to be included? Many years later and for a brief time, I also started to refer to 'my' university as 'we', mostly because I liked and actually admired the rector who led the show back then. When he stepped down, everything reverted to normal.

But more recently, I have started to think that I should stay there on the edge, not let myself get sucked in, retain that distance, because proximity to the frigid sun of the centre blinds more than it illuminates; that I should nurture, not flee, my marginality. But how?

Stand in the Corner

So how about all the others, those whose marginality did not get handed to them on a plate thanks to their childhood and background? Are they doomed to blindness and triviality? Of course not, no less than an upbringing in the margin in any way guarantees insight and perspective. But maybe they have to fight somewhat harder to conquer a little marginal territory, where they are able to genuinely see. In this and the next chapter I will be arguing that it can be good to stand in the corner, on the edge, and that it is both valuable and possible to seek the margins. Many of the examples I will cite come from my own industry, but I believe that the processes are fairly ubiquitous.

What I am trying to say is that marginality is not only a position to which one is born and assigned, an ascriptive property in sociologists' jargon, but can also be something one achieves on the grounds of a strategy of positioning oneself in the periphery and at the points of intersection. An attitude. One can *strive* for marginality just as much as one can strive to be completely engulfed and embraced. One can create one's own marginal position. It costs, but if one thinks a clear eye is worth more than an unconditional embrace, it is worth the price.

In the field of research, the search for marginality is often about finding 'structural holes'. Sociologist Ronald Burt minted this term, by which he means gaps in a social structure, gaps that are not bridged by network contacts but that are, in principle, bridgeable – if only someone knew how and set about doing it. As Burt explains:

> The structural hole between two groups does not mean that people in the groups are unaware of one another. It only means that the people are focused on their own activities such that they do not attend to the activities of people in the other group. Holes are buffers, like an insulator in an electric circuit. People on either side of a structural hole circulate in different flows of information. Structural holes are an opportunity to broker the flow of information between people, and control the projects that bring together people from opposite sides of the hole. [...] Structural holes separate non-redundant sources of information, sources that are more additive than overlapping.[14]

According to Burt, this very brokerage is exactly what an entrepreneur does, an entrepreneur being 'a person who adds value by brokering connections between others. Bringing together separate pieces is the essence of entrepreneurship'.

Such structural holes abound in the social sciences. And I am not necessarily talking of the holes between groups of researchers, but also those between different scholarly literature or methodological traditions. The holes enable us to be knowledge brokers, to add value, by bringing together the hitherto separate.

However, to discover the holes, you must occupy the niche. In this context, the niche can be said to be a form of dual marginality: you are marginal relative to your own research field, the famous colleagues, the familiar; but you are also marginal relative to the others, those who have read other books, who work in another way. If you can position yourself there and retain your place, you can see what others do not see, a tissue of connections spanning what was recently an uncrossable abyss. Just like Neo, the main character in the film *The Matrix*, who chooses the red pill of brutal insight over the blue pill of blissful oblivion, maybe you will find that the scales fall from your own eyes, that you see what other people miss. And just like Neo, you can never be sure that what you see won't fill you with discomfort and fear instead of a sense of calm and security. Maybe you will see things about the society in which you live that you would rather had been left unseen.

Sometimes, the marginal position can really be a matter of life and death, and not just in the world of fiction. The legendary Australian general John Monash was born into triple marginality: in a remote British colony, as a Jew with German roots. He was not a career officer but a reserve officer, and his passage into the officer's mess therefore went not by the traditional route but via national service and then service as an engineer. Monash was thus never schooled in the established military wisdom of the First World War, which was still heavily based on frontal assaults on enemy trenches. He considered this to be an inexcusable waste of human life, letting body upon body be ripped to pieces by machine gunfire from the fortifications. Monash understood that death was an inevitable and dominant aspect of war, but could not accept unnecessary human sacrifice. Initially, opposition among the established British officers' corps to Monash's ideas was formidable. How could this German Jew from the colonies think that he had anything to teach his majesty's finest subjects? And was it really a problem that the rank and file were dying like flies – after all, they were just the common hoi polloi? But Monash refused to acquiesce to the prevailing military strategy and stood firm in the corner in which he placed himself. The infantry, he argued, was to attack under the maximum protection of tanks, aircraft, artillery and machine gun cover in order to save as many young men's lives as possible. This required a completely different form of coordination and more rigorous planning than the then British army practised. With great tactical skill both on the battlefield and in the machinations of the officers' messes, he gradually came to transform the military mindset, steer it onto new paths and end up as the most successful commander of the western front. From his position on the margin.[15]

Although living on the margin, in the niche, does require courage. Even when what is at stake are only arguments and perspectives, not human lives. So much easier to be an indivisible part of the others, so much more secure to live in mutual and constant affirmation, so frightening to forever balance on the edge, to be half stranger to your own kind and half identify with the others. The fear of being ejected, or of falling over the edge, puts many people off going the whole way, and they just stand at a safe distance from the abyss, never discovering what is on the other side.

There are good reasons for this caution. It certainly does not come free, this gaze from the edge, from the fringes of life. Deep scientific creativity is troublingly related to paranoia and psychosis. You are the only one who sees the patterns that others miss, who notes what is going on behind the veil and the masks, how the apparently disparate are actually linked. *The only one who understands how everything's connected.* Brilliance is neighbours with madness, chaos with God.

Remarkably many researchers I know, of a more originally creative bent, have a streak of paranoia in them. They happily smell rats, often feel picked on and persecuted. A handful tip over into mental illness of the kind depicted in *A Beautiful Mind*, the book/film about mathematician and game theorist John Nash and his battle with his psychological demons. Mental illness, when it comes to brain function, has some almost frightening similarities with deep creativity. In such cases, the brain associates sensations and activities that are unconnected to most people, it creates mental chains that are either a sign of brilliance or are thoroughly illusory. 'How could you believe that extraterrestrials are sending you messages?' John Nash's colleague wonders, apropos of his delusions, on visiting the psychiatric institution where Nash is slowly recovering from his disease. 'Because the ideas I had about super-natural beings came to me the same way that my mathematical ideas did,' answers Nash. 'So I took them seriously.' Neuroscientist Nancy Andreasen sums it up: 'Some people see things others cannot, and they are right, and we call them creative geniuses. Some people see things others cannot, and they are wrong, and we call them mentally ill. And some people, like John Nash, are both.'[16]

But even someone who has not strayed so far out onto the margin as Nash unintentionally did can sometimes sense a slight suspicion that things are afoot beyond that which seems to be the case. This very scepticism towards the state of the world is what enables great discoveries and new perspectives but that at the same time abrades everyday life. A hint of the patterns and intentions beyond the seemingly random on the boundaries of existence. *Surely it can't be a coincidence that* [...]*?* is a question that can be a source of groundbreaking insight – or the first sign of madness.

Some of these oblique thinkers can be really quite hard to deal with. They have little need of other people's daily affirmation and do their own thing, sometimes in a rather blunt or awkward fashion. Academic or other leaders, as author Malcolm Gladwell says in an interview, have to protect these wilful individuals from a collective that often finds them an irritant or disruption and that therefore seeks to drum them out.[17] To make everyone show a little indulgence towards their difficult personalities since they are exactly what make them see things others don't. They have placed themselves a little sulkily in the corner, snorting at whoever comes near. But endure them, don't judge them, for soon they might see something others don't see. *If I exorcise my devils well my angels may leave too. When they leave they're so hard to find*, wrote Tom Waits,[18] echoing the sentiments of the many brilliant misfits I have encountered over the years. 'He's so terribly hard to deal with but he's probably the only one at this university who'll ever have a chance of winning the Nobel Prize, so he can be as difficult as he likes. People like that you have to stand up for,' said my former rector over lunch about one of the university's more ornery stellar researchers. Unfortunately, not everyone thinks like this, and many leaders and administrators seem to think it more important to have easily manageable rather than pioneering people in their organisation.

The opposite of conquered marginality is conformism and convention. To some extent they are essential ingredients of the social environment. A functioning society requires a rather large amount of adaptation, convention, habits and routines. If everything was constantly renegotiated and questioned and if everyone constantly spoke their minds, life would be taxing indeed. It is easy to see why many defend themselves against this. But if the conformed and conventional reign alone, everything would stagnate. Norm-breaking and the unconventional are needed to expand the conceivable and question the customary and the taken-for-granted. Being able to see the alternative realities, letting the mind take the leap that frees it from the shackles of established wisdom.

A slightly esoteric question that can be raised here is whether it is the social position one has stepped into that is the decisive factor or simply an attitude that could, in principle, be found anywhere. Is it down to non-conformism and the desire to not really fit in, or to one's vantage point? I believe that in the long run it is difficult to remain in the centre while retaining that open, slightly wondering gaze. That to some degree one must marginalise oneself socially and emotionally too to be able to see really clearly. But I do not wish to rule out the possibility that there are people who manage to maintain an inner distance to their surroundings, to occupy the centre physically but not mentally. I do suspect, however, that there are not that many of them.

But if everyone stands in the corner, the middle of the room will be empty. How can anything be marginal if there is no centre? Is not my proposed strategy self-dissolving? This is a tricky question to which I do not have a good answer. However, I think that the problem will never be about populating the centre. There will always be sufficient consensus-seeking and security-thirsting people who want to be surrounded and embraced by others, by tradition and continuity and the self-evident. Those who have the energy and will to be on the margin will never be so many as to constitute a problem in themselves. And if they ever are, it will be a bridge to cross when we come to it.

In the following chapter I will be discussing what it means to stand in the corner and remain there. How you organise your collaborations, how you relate to your friends, enemies, the collective, the power base. How marginality can be designed and lived. After many years, I have discovered that this, to a large extent, is actually what I have been doing, even though I did not see it like that. That I was continually creating a corner of life where I could be but still not quite be part of.

But the cost of remaining in the margin can sometimes feel troublingly high. I am writing these lines as a visiting researcher at the Max Planck Institute in Cologne, on my own, late in the evening in a darkened office block. It's raining, I miss my wife, I long for home. Or for company, no matter of what kind. At the same time I know that as soon as this no-matter-of-what-kind company presents itself, I will not seek to join in. I will remain on the periphery, joining in without joining. And thinking that I do not really belong there, that I am of another kind. 'I know people whose idea of fun/Is throwing stones in the river in the afternoon sun./Oh let me be as free as them,' writes rock poet Billy Bragg.[19] Which I can relate to, this longing to avoid the feeling of constantly seeking something else, constantly questioning, constantly being on the edge, and possibly on edge. But this feeling is always with me, so deeply rooted in me that even if I wanted to I could do nothing about it. And now it just so happens that I do not even want to.

Being Each Other's Margin

And what should they know of England who only England know?[20] And they who only know their own knowledge do not really know what they know. We need help to be able to occupy and hold a position on the margin. Help from our foes as much as from our friends. In this section I will be concentrating on the friends; the foes will have to wait until the next.

In the previous chapter, I placed the research team in the centre, being the level at which new knowledge is actually created. It might seem strange to go from stressing the importance of cooperation and celebrating the team's mini collective to touting the value of marginality. But in the research team's essential diversity as a perquisite for discovery and insight lies a kind of marginality. A marginality in relation to each other. Being each other's margin.

For this to work, the individual members of your team need to be sufficiently different. Too different, and no contact can be established and you remain mutually excluded; too similar, and you are all ensconced in the middle of the intellectual space. No distance is established, no one learns from anyone else. With Goldilocks diversity, you can be each other's margin; a relational marginality then emerges that is necessary for the whole to be greater than the sum of its parts.

American historian and sociologist Rogers Hollingsworth argues that creativity requires above all the ability to accommodate cognitive complexity; that is to say, the capacity not only to hold more than one thought in the head at the same time but also to simultaneously consider different aspects of a

problem and at that precise moment take in multiple relevant sources of information. This ability, says Hollingsworth, is augmented by team diversity – but
only up to a point. Too different, and we stop bringing anything of value to
the collective table.[21]

There are different ways of creating marginality in the research team. One
is to build it with participants from different scientific disciplines who are at
just the right cognitive distance from each other. What this just right cognitive distance is varies with the problem or the field one is to attack. Often, but
not always, this means that one should work with adjacent disciplines to one's
own. This does not, however, have to be the case. I have seen sociologists in
fruitful collaboration with physicists and computer scientists, and seen them
fail with economists and lawyers. A rule of thumb is that if you recognise
everything that the other is doing, your cognitive distance is too short, and if
what the other is doing appears baffling or meaningless, it is too great. If you
understand the substance of what is being done but have never seen it before,
then the distance is optimal.

An interesting question – which I have borrowed from my good friend and
former collaboration partner Staffan Kumlin – is whether multidisciplinary-
ness is best as input or output; in other words, is it more important and
rewarding to gain impressions from other disciplines that are then reported to
'your own', or to publish for 'the others' on the basis of your own disciplinary
perspective? I am unsure and can easily call to mind examples of good results
from both strategies as well as of failures. Such failures are then not uncommonly grounded in the unwillingness of 'one's own' or 'the others' to take in
perspectives other than the familiar. But they can also derive from a lack of
respect for the others' research from the one who is trying to bring in something new. If you arrive at a new research field and ignore what its researchers
have been dedicating their work to, you risk reinventing the wheel and inciting
their collective anger. Disregarding what others have already done is not a
good way of communicating, and research is very much about communication. Disciplinary arrogance is the cause of many a missed opportunity to
learn from each other.

However, multidisciplinarity is but one of many ways to create the right
amount of team diversity. Another is to put together people of differing
national experiences. Despite all the talk of our globalised world, we are still
largely shaped by the national context of our upbringing. Things we take for
granted and regard as natural are often deeply influenced by national forces,
something that anyone who has moved from one country to another always
discovers. It is particularly valuable for social scientists to exploit such nationally
disparate life experiences. The sociologist who has Sweden as her self-evident,
implicit reference point will benefit from encountering someone whose life

story has other lessons to offer. And is made aware that presumptions are not always to be heeded.

It is not just national backgrounds that create diversity in the team. Gender, class, ethnicity, age, geographic roots: these are all dimensions of variation that can bring vital internal marginality to the team. Monocultures – of whatever form – are bad, especially if one is trying to understand how society works. Even if social variation in no way guarantees intellectual variation – consider the monolithic mindset in the multinational EU Commission – the chances of creating such intellectual variation increase with the breadth of social experience upon which the research team is based.

This does not mean I am prepared to buy a hackneyed argument that what we see is wholly determined by where we stand, that we can never step beyond our immediate experience. 'You will never understand this because you are white/black/a man/a woman/a native/ an immigrant' is an anti-intellectual and constrictive assertion. Traditionally it is usually black people/ women/immigrants who have had this shoved in their faces; today it seems to be at least as common for white people/men/natives to be barred from being anything more than their own experiences. But whichever foot the boot is on, it is a demeaning and intellectually blunting argument that we should dismiss at once. The argument for social variation in the research team is of a different kind – that we can learn from each other and see our own blind spots and shortcomings by meeting other people with other experiences in their baggage. Differences in background and perspective can be communicated and can pave the way for development, provided the circumstances are right.

So if this is to work, we need trust and sympathy. To meet others in the research team on equal terms, to dare admit that there are things we don't understand and can't do, to call for help, is to expose our vulnerability. And this we only dare do with people who we know will not abuse our weaknesses and stick the knife in at the earliest opportunity.

There is therefore good reason to assume that there is some interplay between the diversity inherent to the team and the degree of trust and sympathy as regards how social diversity will affect the team's performance. In environments characterised by low trust, in which people suspect each other of lying in wait to pounce on them, greater social diversity would lead to worse results. In such mistrustful environments, 'the others' will always appear suspect while collaboration and information sharing is confined to 'our own'. The team's members will exclude each other instead of being each other's margin. In trusting environments, on the other hand, the greater the diversity the better the outcomes. Here, information is shared by all, and the broader the experiential base, the better.

Being part of a team also reduces the danger that marginality will turn into (self-)exclusion, that we take ourselves so far out onto the brink that we fall off. The path that leads away from the centre can end in the sect's lair or the hermit's cave. I think this is what sociologist Neil McLaughlin means when talking about 'suboptimal marginality', as opposed to 'optimal marginality', which is the marginality I wish to promote in this book.[22] Suboptimal marginality is more what I here call expulsion or exclusion, a state in which you lack the networks needed to really achieve something and connections with all other actors in your field; in which you do not bridge the structural hole but end up on your own side of it; in which no one knows or cares about what you are doing. Creating and retaining marginality is about maintaining the balance between inside and outside, between being embraced and being ostracised. It is no easy balancing act, and we need our friends to prevent us tipping over, to pull us back when we have gone too close to the edge, to make sure that we are neither banished nor self-excluded.

Maybe it is the case that the optimal team balances and blends not only its cognitive abilities and knowledge-banks but also its personalities. A well-functioning team needs both its strident optimists who believe that everything is possible, and its sceptical pessimists who flag the effectively undoable and ill-conceived. It needs its chatterboxes, who can let words grease a creaking social context and make it permissible for people to talk without having all the solutions at hand. But it also needs its silent observers, who listen and listen before coming up with some timely insight. But exactly like with the purely cognitive distance between the team members, the distance between personalities cannot become so great as to cause aggravation.

There are other things that one might find trying about one's partners when they are duly marginal and complementary. It is, for example, very difficult to state that your friends, perhaps close ones, are wrong, are substandard (this time) and have to rethink and do better. Even though they profess to have already done their utmost. And, conversely, it is just as hard when they think your arguments and results are flawed, that you have to rethink and do better. Even though you profess to have done your utmost.

I have even removed an entire chapter from a joint book at the eleventh hour, since despite all the help and advice the writers had had, it was still not up to scratch. I had to tell good friends whom I had been working with for years that they were not good enough and could not be included. After that we were no longer friends. I think it was something like this that Nietzsche meant when he said that the man of knowledge must learn to hate his friends. It was horrible and I can still feel physical pain when I think about it. But it was necessary. If they had not been jettisoned, we would have all sunk. The least bad is truly not always painless.

This said, it is rare for things to go so far. If our team really does work well together, we will have dealt with our problems long before things reach the point where we have to part as enemies. Like a couple who love each other and refuse to separate despite the irritants of daily coexistence, we will have got to grips with the problems and sorted them out. With sympathy and courage. We will have written and rewritten again and again, been helped over the hurdles, listened and asked and levered everything over the line. Because we trust one another, know that we all mean each other well and that we are doing it together. Because we need and like each other.

But it is not just our friends we need. We also need our enemies.

Look After Your Enemies

I should be more argumentative. But arguing makes me feel so tired and dejected that I often don't bother, even when I'm upset. But even for people who don't like arguing there are good reasons to look after your enemies. There's an expression in Swedish 'to be on edge with each other' (similar to 'to be on the wrong side of someone') that suggests that it is not just the friends in the team who can be each other's margin, but also the enemies, your opponents, the people whose ideas are wrong and stupid, the ones you want to contend with.

The word 'enemy' or 'foe' need not, of course, be taken literally. These are not (or are at least rarely) people whom you hate and want to destroy; they are worthy opponents, people who are worth the effort of defeating or guarding against. Crushing amateurs and imbeciles only gives satisfaction to the sadist. What counts is taking on the opponents in your own event and weight class.

For my part, sociologist Hans Zetterberg (1927–2014) could long serve as the ideal enemy. He researched and commented on my field, was often exciting and verbose without having any proper empirical basis to his theses, and one of few sociologists on the right wing around. I don't think I was ever that interesting for him to mount a defence but he was important for helping me keep my focus and motivation. 'Gotcha!' I almost muttered to myself when my figures could disprove one of his assertions. Even in the absence of an answer I could propel myself onto the next piece of number crunching, the next piece of text. Then Zetterberg grew too old and I found it hard to find a really good opponent in my own field. This is probably one of the reasons why I gradually grew weary and moved over to a completely different one. By this time, the political opposition to my research had also dissipated. Everyone was now claiming to love the welfare state and my figures showing strong popular support for it were embraced by all, from left to right. And when both the scientific and political opposition disappeared, my research suddenly felt rather meaningless and trivial. When enemies retreat, the motivation retreats with them; at least that's what it was like for me.

But from being galvanized by opposition to being genuinely polemic, publicly and cogently, is for me often a giant leap. Arguing sucks the marrow from my bones and when the adrenaline has subsided, all that's left are feelings of meaninglessness, if not hopelessness. Even when I've won the fight. But there are people who never balk at conflict, who even draw energy and meaning from the antagonism.

I talk about this with Bo Rothstein. He's one of those people who rarely shy away from a proper verbal showdown, and it has won him many admirers as well as adversaries. His subject matters are usually well-motivated, even if I don't always agree with him, and you don't really need to look for any psychological explanations for his polemicising. But can he, I wonder a little cautiously, enjoy fighting for its own sake and be energised by it?

'Sure I can,' he exclaims in delight. 'Arguing gives me a kick, a boost, it makes life worth living.' It gives him a lift when life seems grey, conflicts are fun and the polemicising is largely its own reward. 'It makes me feel good,' he says. And he has never experienced any nasty fallout or disastrous personal consequences from his arguing – even if surely there are many who do not appreciate it. He does think it a little silly all this talk about him being 'brave', for what does someone like him have to fear? Those who risk imprisonment and torture when fighting for their and their people's rights are brave, not opinionated Swedish professors who wax polemical. What is strange is that so many other academics are so timorous that they go around thinking that if they lay into the debate too aggressively, they will somehow cause serious harm to themselves. What is the point of professors having career security and academic freedom if they are just going to go around cowering?

Maybe it has something to do with 'Swedes being so conflict-averse'. Conflict aversion is a condition from which Bo, with his background, has never suffered. There were frequent rows in his childhood home with family members giving as good as they got, but it was very accepting and reconciliation was just as quick in coming as rage. 'Shouting at each other was just part of life,' he says. And it is an inheritance he's taken care to nurture.

Yet there is another background to his bloody-mindedness, perhaps, as he thinks, 'a reminiscence of my Jewish heritage'. At Yom Kippur – the Day of Atonement – God goes through the past year, subtracting the bad one has done from the good in a kind of moral bookkeeping. Even though he lost the faith of his childhood years many years ago, this gave him a 'sense that one day there'll be a reckoning', an audit of what one has achieved in life. 'I write for an imaginary panel of auditors,' he says, 'a commission that will question me, come judgement day, about what I have actually said and written.' And from this audit, he hopes to 'come out decent' and not be made to see that there are things that he knew and that upset him but that he kept quiet about. 'It's not so much the grandchildren I worry about,' he smiles. 'It's that commission.'

Interestingly, earlier that same day I heard almost the same thing about the duty of engaging in the debate. The speaker was Olle Häggström, professor of mathematical statistics at Chalmers University in Gothenburg, who drives a usually hard-nosed polemic with his adversaries (especially with those who claim that the threat of climate change is hyperbole) on his blog 'Häggström hävdar' (http ://haggstrom.blogspot.se/) and reviews publications with ruthless candour.

Or as he puts it himself in one of his blog posts: 'I sometimes lay it on the line, without befogging layers of diplomatic cotton wool when faced with people in which stupidity copulates with arrogance, and so I absolutely get it if those who end up in my sights [...] see me as "extremely unpleasant"'. (http://haggstrom.blogspot.se/2014/10/om-trevnad.html) I shudder at the thought of ever coming under fire from his verbal flak, which combines crushing wit with razor-sharp intellect. So far I have not only avoided his darts but have even received consistently appreciative comments, although I'm convinced that the day Olle finds me full of drivel I'll know all about it. Without diplomatic cotton wool.

'I always choose clarity before diplomacy,' says Olle to me in his low-key, intense manner. 'I think that there is generally far too much strategic thinking in the public discourse as it is.' It's seriously detrimental, he says; if no one says what they think, only what they think people want to hear, no one will ultimately understand what the debate is about or what the opinions and desires are of those involved. 'We must be honest, not strategic,' he says, summing up our intellectual responsibility.

And this responsibility is something that hovers over all he does: 'What I have to say is so important,' he says, eyes fixed on me, when our conversation turns to climate change, 'that I feel I have a responsibility to make sure it gets said'. He cites climatologist James Hansen, who when asked if, as a scientist, he should really engage in the debate in the way he does, replies that he doesn't want his grandchildren to say one day 'Grandpa knew which way the wind was blowing but said nothing.'

He has had to endure diverse attacks for his opinions on the climate issue from people who he accuses of being 'climate change deniers'. And Olle often gives tit for tat, often with almost brutal frankness. Unlike Bo, Olle rarely finds joy in polemicising for its own sake. It doesn't energise him, it profoundly wearies him; and sometimes it makes him feel ill at ease – 'But that doesn't make me desist'.

After a while he admits to deriving an almost perverse delight in slating something that is stupid and arrogant, that he can think 'wow, what a gift' when the one he is arguing with makes a real logical error. 'I think in general that it's important to be nice, but when people in an academic position of power act arrogantly and idiotically, I just want to knock them down.' No one could say that Olle Häggström has not succeeded quite well in this ambition.

Another slightly contrary person is Gunilla Gerland, whom we have already met. She often finds polemics 'a cure for boredom'; it's all a matter of 'shaking up the consensus a little, which makes me more alert in my head' and the obvious attraction of engaging in 'tickle-stick behaviour'. Moreover, *not* acting when something seems wrong gives her a worse feeling than the slight discomfort caused by getting on the wrong side of people.

But it's also a matter of how 'opposition makes her smarter', and forces her to sharpen her arguments and rhetoric in a way that would not be possible if she had no one to polemicise against. Borrowing a turn of phrase from author P. C. Jersild, she says that it is like mentally 'going up into the attic' as you manage momentarily to argue at a level that's normally out of reach to you. Thoughts are cleared and words honed in the furnace of opposition.

Gunilla also finds pleasure in never being the one who becomes worked up as the debate heats up. She tells me with great satisfaction about the time she was invited to a TV studio debate on whether single parents should be entitled to assisted insemination. 'I was invited as a "single mother" and promptly decided not to engage in any small talk at all with the others before we went on air, but to just go on and appear benign. Then I just went right in and wiped the floor with them. There's a pleasure in using your verbal gifts like that.' On the other side were two experienced debaters, party leader Göran Hägglund and the then editor and later MP Maria Abrahamsson. 'She is master-suppression technique personified, but when debating she easily becomes emotional and therefore more fatuous. That never happens to me.' And the results were plain to see.

The role and significance of polemics can, as we have seen, be about feeding off the energy of battle. At other times, it is about shouldering what you believe to be an imposed responsibility. Sometimes, it can just be a divertissement or a chance to wise up and hone your arguments. Whatever the case – and whatever the reason for your polemicising – it's all about creating and retaining *meaning* in your endeavour. In the intellectual life, meaninglessness is always breathing down your neck. Why am I writing this? For whom? What does it matter? But if the agenda is clear, the questions' importance evident, the opponent lucid – everything we do has undeniable, obvious meaning. Here is the all-clarifying face-off, here important things are at stake, here prevail concentration and signification.

So looking after your enemies is important, above all, because it is a way of keeping your work meaningful. The worst thing that can happen to the intellectual is not for someone to detest what he or she does, but to not care about it. 'There is nothing so disagreeable as to be hanged in silence', as Voltaire once said. As long as you have someone to argue with, someone who is upset by your arguments and assertions, there is meaning in what you do. When everyone agrees you might as well say nothing at all.

But there can be one more reason why it's important to look after your enemies: they can embody and externalise the nagging doubt and terrifying insecurity that is inherent to intellectual activity. What if I'm wrong? What if I've misunderstood? What if I've missed the point? Such self-criticism and doubt about one's own ability can, somewhat paradoxically, be more easily endured and addressed if it is articulated by someone else. If a person X, Y or Z maintains that I am wrong or am thinking superficially, I can rally to my own defence and show that they are wrong and they my reasoning not just holds but also really brings something to the table. It's so much harder to defend yourself against the inner voice that whispers 'you've misunderstood, your contributions are worthless, no one cares about what you have to say'. Look after your enemies; they can take care of your inner demons and keep them on a tight leash.

Polemicising is therefore a means of saving both the world and yourself when the air becomes too rarefied up in the pure academic search for truth, and when doubt grows about the meaningfulness of what you are doing or about your own abilities. 'In mathematics there are no enemies,' declares Olle Häggström soberly – but I wonder if I hear a slight overtone of regret.

When both self-interest and duty point in the same direction, the path of action appears staked out. After my conversations with Bo, Olle and Gunilla I make a personal vow to argue more. I'm not sure, though, that I'll be able to keep it; maybe I'm just too much of a coward.

Don't Get Drawn In

This is the first time for twenty years I have addressed the House from the back benches.

I must confess I had forgotten how much better the view is from here.[23]

Robin Cook discovered that he could see more clearly once he had left the centre of power, when he was no longer foreign minister and when his gaze came from the UK parliament's periphery. Likewise, any intellectual who wishes to see clearly must not let themselves become drawn into orbit around the power base, become a satellite circling its dazzling sun. This distance is not self-maintaining, as the cool attraction and terrifying force of power is prodigious, in so many ways. I have felt it myself, many times.

I am invited to the Ragnar Söderberg Foundation to talk about one of my books. I naturally accept. When a research patron is interested in listening to what one has say about the organisation and conditions of research, it is an opportunity not to be missed. But I accept for another reason too: for some time now I have been arguing for the establishment of a new commission into power and democracy, or at least a research programme large enough to shed light on the political and economic power shift in Sweden. So much has happened since the previous commission published its report in 1990, so little do we know or understand about the transformation of power and what it implies for Swedish society. But it would be expensive, dreadfully expensive, and who would pay? I pull – very much in vain as it would turn out – every string I can: MPs, research councils, foundations, influential and resource-rich individuals. Maybe the Söderberg family would be willing to contribute? Their cultivational ideals and genuine interest in research is beyond question. And they have resources, financial as well as social, that could be mobilised if they so wished. I want to test their waters a little.

Before the moment, I have some time to chat with Ragnar Söderberg, whose paternal grandfather lent his name to the foundation. The Söderbergs are old money, and in the old communist leader C. H. Hermansson's rundown of the 15 most important financial families of the 1960s, they came just behind the dominant Swedish Wallenbergs.[24] The family has long been established as one of the most important pinnacles in the landscape of economic power. Ragnar Söderberg's handshake is firm, his gaze open: 'I've not had time to read your book yet but you have my word that I shall.' A genuinely nice person, anything else would be inconceivable.

And the presentation is equally pleasant. Around the table are, mostly, representatives of the Söderberg family: a competent little research council is freely assembled from such a family without going outside its confines. The discussion flows easily in the little conference room on the eighth floor with its view of Kungsträdgården in central Stockholm. Vigorous arguments, genuine interest, an amicable but incisive atmosphere. 'I realise that this might be a little provocative for a research foundation to hear,' I say, summing up my litany of the oddities of Swedish research policy. Ragnar Söderberg's gaze sweeps the room: 'I doubt there's anyone around this table who is the slightest bit provoked by your words. Quite the reverse, it's proof that we've been thinking along the right lines.' The combination of such earnest flattery and self-confidence is completely irresistible. I could have stayed for ever in that room, if only time had wanted to stand still.

And I have my opportunity to appeal for a new Swedish commission into power and democracy. There is friendly but slightly cool interest. Maybe the foundation's returns cannot be used for this kind of thing, maybe it is a little too large, and they are unaccustomed (unwilling?) to cooperate directly with other financiers. 'And how does the size of this tally with what you recently said about small teams being the most effective?' wonders Ragnar Söderberg. Excellent question. I swallow and explain that multiple relatively independent teams can work within an overarching framework on different subsidiary questions. And everyone finds it interesting, even if no one wants to make any commitments whatsoever. We part as great pals with mutual expressions of respect, and I still without an answer to the question of how my intended commission is to be financed.

The following summer I make some social excursions into Stockholm's suburbs. I visit Alby and Rinkeby to see how the hard-up live, and to Solsidan and Djursholm to see the physical manifestation of affluence. On a walk along the little deserted streets between the huge estates on the slope down towards the water in Djursholm I occasionally stop to google an address to see who could possibly be dwelling in such residences. And there I spot his name again – Ragnar Söderberg – living in a palatial house right down by the water. A professional landscape architect has obviously designed the elegant garden, and it is equally obviously maintained by staff, as is the house. Beauty and style. A perfect conjunction of residence and resident I think without sarcasm while I trudge on, on the lookout for Stockholm's socio-spatial structure. If, that is, it is the same Ragnar Söderberg – but who else would live here? And the memory of our meeting still feels agreeable.

The glittering attraction of power can be just as strong even when such sympathy is lacking. I recalled a similar feeling of being chosen the time I met George Bush Sr and said 'good morning' to him wearing only a towel. This little incident played out on a paradisiacal, virtually deserted little Greek island. I had travelled there with my family on holiday and we were enjoying the tranquillity. But on this particular morning we could tell straight away that something was going on. There was a racing boat moored to the quay and we noticed some men in white uniforms at our breakfast café talking to the owner.

'Who were they?' we asked once the men had taken off in their boat. 'Bush is here,' he answered excitedly. 'The father, that is, not the son. Out on the big boat.'

We toddled off to our bathing spot, from where we could see the vessel riding at anchor. Before long, a largish party had come ashore and we watched them from a distance disappear up through the village. A moment later, I'd finished my swim, taken off my trunks and was standing with a towel wrapped around my waist. At that moment, the entire party, with George H. W. Bush at its head, came marching past with a spring in their step. Bush raised a hand in greeting: 'Good morning!' I have since wondered what would have happened if I'd dropped my towel or told him what I thought about his son's war in Iraq. It is something I'll never know; a banal human reflex combined with my being star-struck knocked all the things I could have said or done for six. 'Good morning!' I politely replied.

That evening we took our usual walk down to the harbour for dinner. A long table had been set at the restaurant and we concluded that it had to be for the party from the boat. After our meal we stayed behind to see what would happen. One of the racing boats now moored with a single passenger on board. He leapt ashore, shook hands with the restaurant owner and looked around. 'Looks good,' we heard him say, 'but I want a private table and I want it *here*,' he said, gesturing with an officious palm at a place a few meters away from where the big table had been placed. The table was moved to give the man a view of the quay and the harbour entrance. A comms radio was produced and despite our pricked ears we couldn't really hear what was said.

Then another boat arrived, this one carrying five athletically built men. They came ashore and received instructions from the man at the table. 'You three go that way, and you two the other way, reassemble in fifteen.' By this point we had realised that the man at the table was the head of security. The men were soon back reporting what they had seen. Which was clearly nothing untoward, for shortly afterwards a third boat arrived, this one packed with people in casual partywear. The masters of the universe stepped onto the quay and had soon, under much pomp and circumstance, sat themselves at table.

It was like being at a gala premiere. Was George Bush at the heart of the crowd this time too? No, he was nowhere to be seen but we did recognise other people in the party. We soon tired of our celebrity-spotting, however, and decided to go to the pretty little bar at the far end of the promontory and left the group to its fate. The place was deserted, and apart from the bartender we were the only ones there. But now we were here at least, so we ordered a drink each and sat down at the bar.

Suddenly one of the security guards popped his head in. A quick recce of the bar and us, then he was gone. A few minutes later he returned, this time with the whole company from the restaurant in tow. All at once the place was full and the volume all the way to eleven. They completely took over and it wasn't long before some of them had even ventured behind the bar to rummage through the crates of records. New music was put on and soon we were all dancing and drinking together. It was all rather fun and a little surreal – I mean, how often do you get the chance to party with the Bush family? But a glance through the window reminded us of how things stood. Outside, a full security contingent had been posted around the entire building, scanning the surroundings in the balmy darkness. I didn't know if it was the possible threats lurking out there in the dark that were making me nervous or the security guards themselves.

The next morning, normal peace and quiet returned. There was no ship at anchor and all that could be seen was the distant blue Turkish coastline on the horizon. No white boats, no Secret Service, no Bush. We were alone again. And we felt a little bereft; all that energy we'd felt had simply dissipated.

I've thought about this remarkable day over the years. I've thought about the Bush clan, behind the perimeters of their security barrier, where all contact with us mortals is passed through a closely monitored filter. This kind of thing must create a sensation of being chosen, elevated, beyond the limited horizons of mere insects. A sensation that must instil a belief in you that the shape of the world is in your hands. And I've wondered about myself in all this: hand on heart, have my eyes ever glittered so much with such interest as when I was looking at Bush's entourage? Surely it was their very presence that made the evening so magical, wasn't it? Didn't it feel special somehow to be chosen, to be on the right side of the sweeping gaze of the security personnel as they ensured all our safety? Me and my American company in the spotlight, around us just darkness.

Power attracts, even for born sceptics. The fact that researchers are easily drawn to power can partly be explained by the need for funding. As for the intended commission, the implication is that the one with the power is meant to pay for its own scrutiny. A far-fetched idea, perhaps. But there might be individuals among the power bearers who have the will and ability to pay. Indeed, there are private persons for whom an extra outlay of 75–80 million kronor over a five-year period is absolutely bearable. There aren't many of them, but still. Some of them also have a genuine interest in society. As a rule, power and resources go hand in glove, and if you want access to the latter you simply have to interact with the former. The same applies to whoever wants to influence the social discourse and the society in which it takes place: they have to interact with the power base if they are to achieve anything. But take care: it's so easy to slip off the edge, to be devoured by the black hole.

But it's not only about getting your hands on funding or being able to exert influence. The rooms of power are cool and plush, the demeanour is pleasant, the tones dulcet. No one ever needs to raise their voice to get what they want. The civility never feels cosmetic, always natural. Faced with this urbane physiognomy, the visitor easily yields. Everything is so pleasant and who would want to argue with such pleasant people? When they regard you with those amicably curious eyes you feel chosen and eager to remain in the magical aura of their interest in your puny little self. Particularly, perhaps, if your journey there has been a long one.

French sociologist Pierre Bourdieu summed up this phenomenon in his term 'habitus': a series of embodied dispositions through which one relates to oneself and others, a deeply rooted sense of class, which is both an effect of the social world that shapes us, and a social force that shapes this world; that both creates and reflects power.

Social science is particularly vulnerable to being sucked in by power – a phenomenon that is, needless to say, of scant interest to other intellectual activities, such as natural science or medical research. Molecules or microbes do not return the scrutiny, they think nothing of what the researcher is claiming about them, and they have no active bearing on the researcher's future.

The danger of being sucked in is greatest for those whose subject lies in close proximity to the power holders, who actually study political and economic power. How often do we not see this? Social scientists who allow themselves to be reduced to sports commentators when the latest political manoeuvring is to be commented upon and its effects on the electorate discussed. Or to a royal advisor when the prime minister or other power bearer is to be informed about how best to handle a particular situation. I feel sick every time I witness this. It's horrible to see how easily scientific independence is abandoned before the opportunity to dance in the magic circle of power, in the glittering footlights of the mass media, in the court of the prince.

Because it is this lack of independence that disturbs me. This relinquishing of the agenda to someone else. The problem is not being part of contemporary life and following its currents and participating in its conflicts and debates. This is expected of us social scientists and rightly so. The ivory tower is not a place where creativity flows. It's just this, letting those one is to study – the power holders – decide the nature of this study. That one so obsequiously plays along instead of presenting one's own agenda, one that breaks with the presupposed and established instead of rubbing it up the right way.

The focus on 'partnership' which characterises research policy around Europe therefore poses a direct threat to the status and quality of social science. Partnership in the social sciences often entails either marketing innovations with private companies or (more commonly) helping to solve immediate organisational problems for non-academic actors, who will (usually) be the ones who formulate the very problem or problems to be solved and, at worst, the solutions that should be adopted. Thus is social science reduced to the handmaiden of power, trapped in the individual perspectives of the parties involved, unable to uphold the independence and distance without which no social scientific research worth its name can exist.

In the case of 'partnership', lack of independence emerges as an unfortunate side-effect of the researchers being drawn, deliberately or not, into an unbalanced collaborative partnership with others. In other cases, the encounter with power is even more unpleasant as it actively seeks to erase the independence of the researcher and his research. Political scientist Sverker Gustavsson recounts his early days as undersecretary of state at the Ministry of Education, how both independent consultants and the finance ministry's own civil servants swiftly approached him with the intent to 'rationalise' the unruly academic world. Abolished employment security for researchers – they were to be on a three-year contract at a time – and the consolidation of all power to the university management as regards when, how and by whom research was to be conducted were two of the ingredients intended to ensure a more compliant research community.[25] Gustavsson was having none of it and the most radical proposals were never – or not yet? – implemented. But scratch the surface and you'll find that the desire to control research and its practitioners remains.

The view of the role played by the social sciences in societal development and in the general social conversation is often marked by a certain naivety. It is not uncommonly depicted as 'research information', a reporting of proven and established results that are to inform enlightened and well-considered political decisions. But this is not what the social science contribution actually looks like, and this is not how politics actually works. The results of social science research land in a conflicted political life, where they are often used as political ammunition against opponents rather than for decision-making in the public interest. Social science research, when at its best and left sufficiently to its own devices, can provide vital input to an animated conversation about what our societies look like and how they should be constructed. But not everyone wants to hear all the stories and some have the power to silence their teller when their moral is inconvenient. Or just severely stymie the research with their intransigence, by refusing to be interviewed, withholding information, undermining trust in the researchers and their research.

The hot air about the importance of independent research is sometimes just this. I would venture that this is partly because social science is in a kind of competitive situation with political actors, public authorities, mass media and others as regards defining, describing and explaining societal phenomena. Power holders are therefore always driven to rid themselves of these troublesome social scientists, to be left in peace with their own attempts to define reality and not to be constantly confronted with awkward questions about evidence and relevance; to shut the researchers up once and for all.

Often they keep this impulse in check – stopping or controlling independent research certainly awards them no brownie points in the public discourse – but the impulse is there nonetheless and motivates different kinds of interference with the conditions and opportunities of social research. The social scientist must constantly fight against being sucked in, against being eradicated for good.

However, not being drawn in, keeping your distance, is not only about steering clear of the gravitational centre of power. It's also about – and this is perhaps even harder – keeping your distance from your friends. Distance from the self-evident, from the given presuppositions we eagerly cultivate in company with others. Where we validate each other, smooth over and seek consensus, cloud each other's eyes.

This constant seeking and testing, this stubborn questioning of the established, is often hard. Hard for those who constantly have their presumptions questioned, but often also for those who can't help questioning. The party pooper, the one who says that tedious 'it's actually not that simple, there are just as many counter-instances' when someone blithely throws out an unfounded statement or a half-true prejudice. 'The best thing about leaving the sociology department was being able at last to sit in the coffee break and have a natter without constantly being questioned and called upon to produce evidence,' a former colleague once said. I have no difficulty understanding that sentiment even if I don't share it.

It can feel so snug, being part of that self-evident company in the middle of the room, where everyone agrees and mutually affirms. And how many times have I myself not been bothered to break free, when I let the conversation revolve around the 10 per cent of the stuff we agree on, and not the 90 per cent we don't. Humans are pack animals, deeply dependent on others for their survival. In this dependency nestles the solid core of conformity, the ostensibly amiable adaptability to the majority. Being abandoned, during our delicate infancy both as individuals and as a species, meant *death* in a very immediate and literal sense. It is not so strange, then, that we are so fearful of abandonment and so keen to say things in an inhibitive quest for accord. Even if actual physical death rarely threatens the non-conformist in a rich and pluralistic society, we often fear social death. We fear being the one that others go out of the way to avoid, the one who's talked about in whispered tones as soon as they've left the room. The creative margin must be constantly re-captured from our yearning to be fully included, from our desire to let our ideas get drawn into the opinion of the collective, from our fear of abandonment.

Movements in Space

Of course I exaggerate when I say it, but if often seems as if my entire life has been one long journey from the periphery into the centre. While there has been some back-and-forthing between Umeå and Stockholm and London, ending up ultimately – after a long time – in Umeå, the journey, on the whole and both geographically and socially, has tended towards the centre. From the little yellow teachers' home in Jockfall to the ten-fold bigger Överkalix, via the university city of Umeå to the capital Stockholm. From the forestry village to the Swedish Research Council and the Royal Swedish Academy of Sciences. From social outsider to the heart of higher education.

Even if my movements have largely been in a wholly Swedish national space, I think that I share the experience with the many migrants who have made the journey from the periphery to the centre the world over. From the rural American South to the cities of the north, from the villages of Africa and China to the metropolises. A migration from the rural to the urban, from the familiar to the strange, from the conventional and rarely questioned to the negotiated and fluid.

Most movements of this kind are driven by necessity and privation and by a feeling that things at least can't be as bad as they already are. And many of the journeys end in disaster, with corpses washed ashore on the beaches of Europe or riddled with bullets in the favelas of the poor. At the same time, however, movement possesses incredible potential for people to expand their territory and resources, and to see the world in a new light. Perhaps in ways that no one else has thought of because they don't share your journey and what you've made of it.

The migrant is by definition a marginal person, someone who will never again be fully part of what they have left behind. Myself, I knew that with each of the moves of my childhood and teens, I would never return. For a brief visit, of course, but not in any permanent sense. And even for those who do actually return, the place is never really quite the same. It's impossible to answer the call of nostalgia. Something has happened, not so much to the place as to the one returning to it.

But by the same token, the migrant is never part of their destination. Her experiences are too remote from those who have never left their environment. She has memories and perspectives others lack, and things that the others take for granted she does not. She wonders, perhaps constantly, why the natives seem to be so unquestioningly anchored in what she knows looks different from another perspective. Why they don't seem to see the water in which they are swimming.

A glance at Europe's cultural modernity reveals that it has often been borne aloft by those who have moved from the periphery to the centre.[26] The archetype here is perhaps Pablo Picasso, and his movement from Malaga, the town of his birth, via Barcelona, to Paris and the hub of the artistic world, where he arrives in 1900, not yet 20 years old. What Picasso had with him from his childhood home and the never-completed art schools of his youth are transformed in his encounters with other Parisian artists and the wider cultural circle in which he now roams. Naturally, the circle contains artists such as his great rival Henri Matisse (possibly the only one whom Picasso acknowledged as his equal), his close partner Georges Braque (who launched with Picasso the movement that became known as cubism) and countryman Juan Gris (another luminary of early cubism). But there were also poets like André Breton, Jean Cocteau and Guillaume Apollinaire. All of them had moved from the provinces to the centre, all came bearing experiences that to some extent, but not wholly, were shared by the others. It was a crucible, or perhaps a witches' cauldron, of ideas, perspectives and techniques.

All this was brought forth in and through Picasso's artistry in a way that changed art history. Picasso devoured life – part man, part beast he was. He constantly let down his friends and family, and his women and children played a horrifically high price for the ruthless monomaniacal creativity that reformed the art of the present and the future.[27] *Bearing* the cost of marginality can seem valiant; *burdening* your nearest and dearest with it is repugnant.

As famous is an earlier migration from the margin that changed art history: Vincent van Gogh's, from the little Belgian town where he grew up to the heart of the artistic world in Paris. Like Picasso, with his baggage of having begun but never concluded art studies. Marginal in relation to childhood environment, destination, education, the lot. And, eventually, the ultimate borderline experience, the psychosis, the all-consuming mental illness. The nervy explosion of colour that is his painting was born from borderline experiences to which few have the courage and ability to expose themselves. From the margins of madness and death comes a gaze that sees what other people don't.

The Parisian artistic circles of the late nineteenth and early twentieth centuries were certainly no idyll. Most artists had great difficulty making ends meet, and many lived in abject poverty. And the cliques flourished as much as the antagonisms. Picasso's rivalry with Matisse is one of the better known examples, as is van Gogh's row with Paul Gaugin, the aftermath of which led to van Gogh cutting off an ear. Blood, tears, harrowing rivalries and score-settlings. And in the midst of it all a flowing creativity of mutual learning, between friends and enemies alike.

I think about the father-in-law I never got to meet, artist Stig Sundin (1922–1990), who passed away 20 years before I met his daughter. Raised in Hammarstrand in rural Jämtland county, born out of wedlock and therefore discriminated by the family even though his parents married shortly after his arrival. At the age of 12 he took a job to bring some money into the home. Meanwhile his dreams of being an artist grew, and he began to secretly save; before long, he was able to buy a bike and embarked on the ride that would take him to Stockholm and, eventually, the Royal Academy of Fine Arts and the life of an artist. What did Stig see that the others didn't, those who had been born in the centre of art? This is something I will never know, as he wasn't one for theoretical reflections upon his life and work. He would often snort at the mention of the word 'creativity', since he associated it with lazy inspiration, fully aware that it was only hard work that gave results. But I imagine that his gaze was different, that his journey into the centre from the periphery gave him a perspective that others lacked. 'You can't *become* anything, you can only find out who you *are*, and then *be* that person' was the harsh message he gave to his daughter as she sought his opinion on what she should become. A piece of advice as good as any for all seeking, searching people.

The migrant can have moved from one geographical area to another; but she can also have moved socially, with or without any spatial relocation. Most such talked about journeys pass upwards in the hierarchy, taking formerly excluded or subordinated individuals and groups over the threshold into the finer salons and the centres of power. The opposite transition is less often mentioned, when people sink and lose what they once had.

The academic footing gained by the feminist movement, the broadening of the universities' recruitment base, the 'moving up' of individuals from a class of privation to one of affluence: all these are examples of journeys in the social space. And they all transform their destinations as much as their travellers. For example, the social sciences have been dramatically reshaped by the impact of gender studies. From having been focused on the situation of women and with a feministic will to change in the back- or foreground, gender studies today concentrates generally on how gender both constructs and is constructed by societal institutions and human lives. New questions are raised and new perspective applied by those who travel in from the margin to the centre.

But above all, by those who never quite make it. Those who step body and soul into the centre consolidate rather than transform this centre. Like Ove Guldberg, the monstrous defender of power and court whom Per Olov Enquist depicts in his novel *Livläkarens besök*.[28] So unlike his enemy, Enlightenment man and royal physician Struensee. He loses the power struggle in and around the Danish court as well as his life, but perhaps sows the seeds of the tree whose roots will eventually overthrow the autocracy and establish democracy. Struensee never quite makes it, never fully learns to play the court's power games, a skill that could have saved his life. But maybe he retains, by virtue of this ignorance, something of the fresh gaze that comes from the margin, from the one who travels and travels but never quite arrives.

The geographical journey does not, of course, necessarily have to lead towards the centre – it can also head in the opposite direction, from the centre to the periphery. 'Perhaps genius needs to gain distance from wherever it first resided,' muses Howard Gardner in his book on creative transformers of culture.[29] He observes that while most creators moved from the periphery inwards, in some cases – poet T. S. Eliot being his most notable example – the journey went in the opposite direction.

Newcomers in the periphery will discover, then, that they will never really be included in their new context. Just like my family and I never really became part the Jockfall community, the new arrival never really becomes part of the periphery's centre. And in these contexts, 'newcomers' are often considered so in terms of generations rather than years, although this rarely means that they become ostracised or outcast. They find their place in the periphery's marginality and often contribute to renewing their destination. Only by coming from elsewhere, by having seen and experienced things other than the forever rooted. By existing in the margin.

And then new people embark on new journeys into the centre, often to escape what they see as a stifling childhood environment. With every step in towards the centre, I myself have experienced a widening of horizons, have found people to talk to and be with that were not present in the places I have left. People who, like me, have thought a little more than many others about how everything worked, why things were as they were, and what can be done about it. These were rarely the kinds of thought one would have had in the places in which I grew up.

A marginality that does not contain any movement in space often risks ending in self-exclusion. The local eccentric, the hermit in the cave, the village idiot: for some of them, the stationary journey out into isolation could have ended differently if they had only been able to embark on a real journey, one that took them to new horizons and perhaps new acknowledgement of a kind they could never have in the intellectually impoverished soil that was, to them, a given.

Author Mikael Niemi concluded his appearance on Swedish Radio's summer talk show with an emotional appeal: Northerners in the diaspora, come home, we need you! For a moment I find myself suddenly standing there in the middle of the kitchen, ready to heed the call. But the next second I've got my wits together. What would I do in Överkalix? Who would I be able to talk to about the things that mean something to me? How strange I would feel, again.

We gotta get out of this place, if it's the last thing we ever do, sang The Animals, and how many times have I not hummed along with them? No, I don't want to go back to where I once belonged, no matter how happy I am that I originate from there. I raise my eyes from the screen, look out at a German city I barely know, and am glad that it's so unfamiliar, that life is waiting out there to be discovered.

Chapter 3

DARKNESS AND LIGHT

Luke: *There's something not right here [...] I feel cold. Death.*
Yoda: *[points to a cave opening beneath a large tree] That place [...] is strong with the dark side of the Force. A domain of evil it is. In you must go.*
Luke: *What's in there?*
Yoda: *Only what you take with you.*[1]

The Heart of Darkness

We were used to being pushed around and losing, but this time it was if we were possessed by the Devil himself. The boy was a couple of years older than us and normally he just took what he wanted, in this case the new football we were playing with. But this day was different. I don't know why – perhaps we just snapped. When he tried to take the ball, we pounced on him like enraged furies, brought him down onto the grass, punched and tore and ripped and soon his new white trousers were covered in grass stains. He stared in horror at the mess, started to sob and limped away. We'd won. And we knew that it wasn't our surprise attack that had made him cry. It was knowing that his ruined trousers would earn him more than his usual beating when his father came home. We all knew that his father beat him, but no one said anything. It wasn't a village where you stuck your nose into other families' business.

Of course, this insubordination had to be punished. A few days later he got hold of us in the playground, and led us to a hidden-away corner where his mates held us fast while he punched and punched. We returned after the break, all teary-eyed and roughed up. Naturally we said nothing to any adult – it just wasn't something you did – but it was obvious that our teacher had got the wrong end of the stick and thought that my friend and I had been fighting each other. How I would have loved to tell her that that wasn't the case, but the playground culture of silence stopped me. Despite my pain and humiliation, I took tacit pleasure at the thought of the beating he must have been given when he got home, which was probably much worse than he'd given us. The pecking order had been re-established, but that intoxicating feeling that revolt was possible stayed with me.

This part of the book focuses on the darker sides of research and writing. I will contend that what we usually think of as negative emotions to be worked through or avoid are not only inevitable but also indispensable for researchers and writers. Without a trace or even large dose of black emotion – such as fear, anger, angst, tedium, self-contempt – there can be no dark background against which the text can shine. The darker the shadow, the brighter the light. But if the fear, anger, angst, tedium or self-contempt takes over, we are crushed. On this knife-edge of emotions balance science and the intellectual life. We need all the help we can get if we're to make it.

There's a genre of pop-psychology management book that preaches the importance of 'daring to fail';[2] these books are often written by and for people who in actual fact have little to lose by failing. The safety net is tied nice and tight for when they slip on the trapeze. Back on terra firma they congratulate themselves on their courage, convinced that this very fearlessness is the secret of their success. Safe in the knowledge of future success, they boldly fail again and again until they finally manage to do whatever it is they have undertaken to do. I do not wish to find myself in such smug company. I want my story to have much darker tones, to be one that brings no guarantees, has no instructions, and that takes courage in a much deeper sense to enact.

I have lived for so long in a constant state of failophobia. I overrun no deadlines. I don't take on too much. I always deliver what I've promised. I usually arrive too early, never late. Many would see these as strengths, but it's more complex than that. I'm not exploiting my full potential. I'm not using all the time I have at my disposal. I'm not reaching as high as I can go. I'm simply too faint-hearted. So when I argue that we should have the courage to approach the brink of the pit, I have to confess that I myself have rarely dared to stand there. I want others to be stronger than me. Maybe somehow that would make me a little stronger too.

I think that it all comes down to the fear of getting sucked back once you've 'moved up' a class, being gobbled up again by the world you've broken free from. Like a shark that has to keep swimming in order to breathe, once you've surfaced you have to constantly work at not sinking again, all the time with a nagging worry that you might not cut it, that they might send you back from whence you came. So I'm constantly having to prove my own worth, not so much to others as to myself. And this means no failures.

It might therefore seem strange that I am voluntarily tackling this theme, or that I chose research as my profession in the first place. Research is – as I shall return to – really an institutionalising of failure. Criticism and self-criticism are built into its very core, the setbacks greatly outnumber the advances, and the elbow room is confined and constantly under siege from rivals. As I start to write this chapter, I haven't been given any of the last three appointments I've applied for, have had three out of four grant applications turned down, and have had my three latest essays refused by different journals. Nothing deserving a Facebook post there. Yet these refusals and rejections are completely normal and an everyday aspect of my chosen field. They're something we can choose how we deal with, but nothing we can choose to eliminate.

Research into research has not been particularly inclined to take the researchers' emotions seriously. Admittedly, in my own scientific discipline of sociology research on emotions has blossomed in recent decades and is now a vast, complicated field.[3] The social plasticity of emotions, their implications for power and stratification, and their influence on our social behaviour are just some of the subjects into which social researchers delve.

They rarely, however, interrogate the significance of the emotions for research and its practitioners, nor is there much literature about it beyond the sociology shelves.[4] It is as if the researchers prefer not to talk about such matters, perhaps out of fear that it will undermine credibility in their research or because someone had convinced them that emotions are merely a disruption. Despite the fact that scholarly autobiographies often stress the part played by the emotions in scientific discoveries, the principle arguments are often about how emotions are best avoided in the interests of good research. But in actual fact, emotions are 'an unavoidable and essential aspect of science in all of its phases, and underlying the social basis of science itself,' according to one of the few papers I've been able to find that takes the role of the emotions in science seriously.[5]

On those rare occasions that the role of emotions in science is discussed, the focus is almost exclusively on the *positive* emotions. The euphoria of discovery, the joy of collaboration, the pleasure of curiosity; these crop up here and there in the odd text.[6] But pretty much nothing about the significance of fear, anger or tedium. It seems implied that such things are to be avoided. We must learn to be less afraid and angry and not to get bored by what we are doing. The black emotions are to be suppressed lest clarity of thought be clouded. I think, on the contrary, that these black emotions have an important role to play for us in our knowledge-seeking and I would like to attempt an explanation of how and why.

Taking the researcher's emotions seriously is to say something about what it is to be human – in this case a human who expresses herself in the form of a text on the structures and changes of society. Too often, the researcher is depicted as 'a strange animal who eats money and shits knowledge' to quote author Henrik Berggren on the apparent view of the then Swedish minister of education.[7] But we are human beings and from this self-evident fact follows many things that neither we researchers nor those who finance or try to control us are particularly disposed to talk about. We are elevated and tortured by our emotions; we are not brains on legs, we have bodies and joy and anxiety. Acts of wickedness leave their mark as much as those of solicitude. But we turn a blind eye to them, preferring to tell tales of the researcher and her research, while the real history gets left behind in the shadows. I try to expose these shadowy things to the sun so that the trolls that populate them may shatter in its glare.

I am talking here of the hardest things an intellectual can do: take on his own shadow, handle and make productive use of his soul's dark side. No one unable to manage this will ever, in the long run, be able to be intellectually active and will fade in his flight from the pain and fear, or be crushed under their enormous weight. Their shadow constantly falls on he who seeks knowledge and literary conceptions. Pressed in the right order, the keys in front of me could create the world's greatest novel, or a revolutionary research paper. But in what order? And who is the right person to find it? Not you, says the shadow.

Some social scientists are far too kind. They only think well and are inclined to don rose-tinted spectacles. No wickedness inheres in them, no black glint is seen in their eyes. And their texts never sparkle.

Fear, anger, tedium, angst, self-contempt. This is a journey into the heart of darkness.[8] But the one I meet in the foreboding depths of the jungle is no Colonel Kurtz. It's me, waiting to see if I can make it all the way.

Being Just Scared Enough

Many people like to idyllise their childhoods. They remember it, or at least like to depict it, as a carefree time before the burdens and responsibilities of adult life were laid on their shoulders. In their memories, their childhoods were constantly happy and a little naughty. But that's not how I remember mine, even though it wasn't at all disadvantaged or unhappy. My clearest memory is one of frequent fear.

I was afraid of the big boys who smoked and drank beer and rode souped-up mopeds. They rarely gave me any trouble, in their world I was completely uninteresting, but their mere presence in the playground at dusk felt intimidating.

I was scared of the big waterfall in our village. My mum had imprinted a little too carefully in me how dangerous it was and when I stood on the rocks by the surging torrent, I knew that death was lurking down below. I often had the sensation that in some strange way I'd stumble and vanish forever in the swirling waters.

I was scared of being made to do something I didn't like or simply found offensive. I remember, for example, how I was mildly coerced to go on sadistic vole hunts. Armed with croquet mallets, the boys went down to the meadow, located the vole holes, lit a fire at one exit and stood ready to strike at the other. Eventually, the vole was smoked out and *biff*, the mallet fell and the vole was turned into a bloody mush. Sometimes they had to hit the poor thing several times before it stopped twitching. I hated it, and didn't want to join in; I can't actually remember, though, if they made me strike the fatal blow.

But I was especially scared of being late. It almost never happened as our garden bordered the school playground and it wasn't more than 50 metres door to door. Despite this I was often first in line outside the classroom. The idea of arriving late, seeing everyone's eyes turn to me as I walked into the classroom, getting told off by the teacher, having to explain in front of the others why I was late, was unbearable. Hence the early arrival, hence the panic on the rare occasions that I overslept. I'd rather have bunked the whole day than arrive five minutes late.

The scared often arrive too early or not at all.

Maybe the art of arriving in time is the most important thing for a scholar. It can be conceived of as a matter of how precocious one should be. Not precocious enough and everyone is saying at the same time what you thought was a unique insight. Too precocious, and no one will understand you. The acknowledgement of posterity will then – at best – compensate for the dismissal of the present. Compensation that you will never enjoy.

But arriving in time is also about being ready on time with whatever it is you've undertaken to do. It's a matter of being able to navigate between the Scylla of procrastination and the Charybdis of precrastination. Of not coming too late, but not coming too early either. And here, fear, especially the fear of failure, plays a fundamental role.

Putting things off until later – procrastinating – seems to be the faithful companion of the academic life. Many scholars seem to regard a deadline almost as a suggestion: 'It'd be good if you could be done by then.' For the journalistically schooled, this is a joke. They know that anything that doesn't make the deadline will just end up in the wastepaper bin. But academics love putting things off until the last second, or even later – which is particularly irritating for the minority of us researchers who are never late.

Others – and I confess to being one of them – spend their time *pre*crastinating. We get things done too early. But what's the problem, an inveterate latecomer might well ask, being done too early can't possibly be a problem in itself? Sure it can, says Oliver Burkeman in a column in the *Guardian*.[9] The pre-crastinator hurries to get things done, ticks them off on his list so that he can feel calm and at ease. The word was minted by some psychologists, who found that their study participants, when asked to carry one of two heavy buckets over a finishing line, often chose the one placed closer to them instead of the one placed closer to the line.[10] They simply wanted to get to grips with the task as soon as possible, and so the closest bucket seemed the right choice.

The pre-crastinator always finishes things too early and therefore doesn't use the time at his disposal, doesn't work as close to the deadline as possible, doesn't stretch himself. The pre-crastinator is often inefficient, as he often has to revise the work he's so diligently ticked off his list before delivery. At worst, his eagerness to be ready with things steals his focus from the long-term goal: the to-do list of less important chores is ticked off, while the thing on which he should really be spending his time gets pushed into the background.

How can highly educated and intelligent people behave so irrationally? It's fear. The fear of failure that lies at the root of both the notorious procrastinator and the incorrigible pre-crastinator.

For the procrastinator, it's about accepting or abdicating responsibility for her own thoughts. It's not until these thoughts are committed to print that the author becomes responsible for them, is made to acknowledge to herself and others that this is what she wants to have said. And maybe it's wrong or superficial. So it's not so strange that many people avoid this anxious moment for as long as possible. Particularly if the deadline is more of a suggestion than anything else.

For the pre-crastinator, it's all about the fear of failing to be done in time. This is something I've experienced many times. By nature a serious time-pessimist – why would I otherwise have allowed 15 minutes to walk the 50 metres to the classroom door – I always think that I have less time at my disposal than is actually the case. Relieved, I eventually realise that I've actually finished what I'm meant to be doing long before the allocated time runs out. Then I often have to remind myself of what that lecture I prepared far too early is actually about when the time comes to hold it.

I've never been able to understand those who time and again keep others waiting. If it were me, I'd die of shame. I'd be standing there at the classroom door again, watching all the faces turn to me. But I think that the constant latecomer finds it just as hard to understand my behaviour, and no wonder – I can myself see that my inability to learn from my mistakes isn't that easy to comprehend.

But the fear of failure not only compels us to act irrationally, it is also behind many of our successes. It makes us go that extra mile, stretch to reach the unreachable. Strive upwards along that infinite yardstick. If we ever achieved our aim, we'd have to find something else to reach for.

What about, for example, that sudden creativity that often strikes right at the end of a long, unproductive day just before the kids need picking up or dinner needs to be made? Is that also not a positive effect of the fear of failure? A desire to not to have to meet yourself when darkness falls and admit, in front of your most ruthless critic, that your day has produced absolutely nothing. Hence the spurt just before the finishing line, hence the vain calls for more time. Futile. Because if there had been more time, it would also have been frittered away and only the final minutes of the dying working day would have been used. When Hegel stated that the owl of Minerva spreads its wings only with the falling of the dusk, he possibly wasn't only referring to the understanding of history but also to his own work rhythms.

The fear of failure is intimately linked to what in the first chapter – with the help of sociologist Dalton Conley – was described as *fraud anxiety*.[11] Psychologist and organisational consultant Valerie Young speaks of the same thing with her *impostor syndrome*, that nagging or even acute feeling of being a bogus intellectual.[12] Young's book is for all the women who feel frauds, who live with the constant feeling that 'they made a mistake in employing/promoting/listening to me and I'll soon be rumbled!' Even though similar anxieties can plague men – particularly those from the margin – Young is probably right in saying that they are much more common and crippling in women.

However, I'm not so convinced that Young's suggested solutions are correct. They forever roam along an individualistic track, explaining how the addressed 'you' can change 'your' actions and mindset. Useful advice, to be sure, but really the book ought to target the environment which the self-pronounced imposter occupies. Because I believe it is only the social relations we create and maintain that can alleviate our sense of fear and of fraudulence. Being surrounded by friendly and honest people who genuinely care about what we are doing. Perhaps Young thinks that this is nothing that the sufferer can achieve unaided so it is better to focus on personal strategies. But sometimes, and to a certain extent, we can actually choose our immediate environment and perhaps this is the most important step we can take to deflate the fear to a manageable level and mitigate the nasty feeling of being nothing but a fraud.

Or as Malcolm Gladwell writes, describing how important an accepting and forgiving environment is to preventing the feelings of fear and fraudulence taking over:

> The most powerful weapon against fear is forgiveness. If you are part of a community or a context or a world that is comfortable with the idea that people are sometimes fearful, sometimes make terrible decisions,

and sometimes don't do what they are supposed to do – and you continue to support them – then it becomes a lot easier to overcome fear. The key to overcoming fears is your understanding of what happens after you have done or not done something – and if you know that what happens next is that you will continue to be supported, that makes it easier to do the right thing. I think of things not in terms of the individual but of what surrounds the individual.[13]

A TV show that I can't now find talked about music therapist Louise Montello (1956–2014), who helped young musicians who'd felt that they'd reached a dead end or even that their music had 'died'. Technically speaking they were perfect, but their music didn't sound so good and they were suffering both physically and mentally because of it. Her approach was to encourage them not to be perfect. Most of them had been drilled ever since infancy, their natural talent having been developed and honed until nothing but perfection remained.

One of them plays a piece. Brilliantly, impeccably. Then Montello plays the same piece. Even I – with my fairly deaf musical ear – can hear how much better it sounds, so much more vibrant, despite her not having the same level of technical acumen as her young disciple. It's all about unloosening one's true self even if it means making mistakes, she instructs, and the one who's had it drummed into them throughout their life that mistakes must be avoided at all costs listens attentively and starts to slowly acquire their own way of approaching music and the world.

Montello says that these musicians are suffering from an 'uncompromising perfectionism' that is holding them back both as musicians and as people. She continues:

> This extreme form of perfectionism seems to be rooted in the vulnerable musician's need for outside approval and validation of his or her innate worth. When these musicians receive unfavorable feedback associated with a musical activity, they often experience it as a reflection of their inherent 'badness' and judge themselves as being deficient. Because many musicians' self esteem is dependent on how others perceive them, they often lose touch with their own inner reality – the essential self – which holds the individual's unique life purpose, innate gifts, feelings, beliefs and somatic states. Instead of trying to connect with their own essential selves, these musicians try to imitate the process of others whom they deem to be 'successful'. Vulnerable musicians often are unable to accept their own humanity – which is, of course, prone to error, fatigue, fear and other less than ideal self-states.[14]

There is evidently much at stake here. People's sense of self-esteem, their creativity, the acceptance of their own humanity. So many of these apparently perfect yet rather fragile beings I encounter in the world of research. Obsessed with *doing things right* instead of *doing the right things*, they're often possessed of a technical brilliance superior to most others'. But their research is never really urgent, stymied as it is by their desire to never do anything wrong, to never suffer the ignominy of censure. Sometimes they become less fearful as time passes and if they are still able to retain their sensitivity, it's then that their glory days come. But sometimes their conquered fear turns into complacency and mental dullness and they end their intellectual days in crotchety retirement.

There's a video clip of legendary graphic designer Milton Glaser talking about the fear of failure and what it does to us.[15] Glaser argues that what constitutes professional success – finding something unique and then repeating it – is often the antithesis to personal development. Personal development is about heading towards failure, failing, learning from the failure and then growing from the experience. In this movement towards the threshold of failure we are hampered by our fear. We might have to endure with the judgement of others, being criticised for our failures is no fun. What's more important, however, is the difficulty of handling self-criticism, as being forced to admit to ourselves that 'you're not as good as you want to be and as others think you are'. There is only one way out of this fear, says Glaser, to 'embrace failure', to turn it into your friend, into a step in your personal growth.

Wise words, but easier said than done. There is today, as I've mentioned, a whole genre of notions about how important it is to dare to fail – usually written and promulgated by people like Glaser who have themselves succeeded, who have, indeed, become almost iconic in their own domain.

Fear can be the driver of extraordinary achievements. Going the extra mile is maybe what only the rather fearful do. Actor Stellan Skarsgård talks here about his fear as being both engine and brake:

'If you're at the pinnacle of your theatre career, the only natural state is fear.'

You mean that fear is a motivator?

'No, although possibly what drives you is the avoidance of fear. The fight against fear can galvanise you. [...] Beware the actor who isn't scared,' he suddenly declares, adding with a low chuckle: 'They love themselves too much. [...] To create something good you need to be completely open and pretty damn fragile, somehow. The higher your ambitions, the closer to that brink you have to go. If it doesn't work, a black cloud of panic blows in and you can't remember your lines, you can't move and speak at the same time. You just want to die.[16]

You just want to die. Maybe it's the fear of fear itself that deters so many people, like me, from getting really close to the edge. How will we muster the courage to go further? *Everything comes to those who dare to be afraid*, sings the Swedish rock group Kent, but dare I heed the words? All I know is that I can't do it alone.

Violinist Joel Sundin tells me about the career that was never chosen. His engagements are global: while he has often been based in Copenhagen with the world-famous baroque ensemble Concerto Copenhagen, his career has taken him around the world in the company of other brilliant musicians. But it's still something that he maybe wishes he had done differently:

> *Have you closed off your roads and doors because you've been too timid?*
> You could say that, I suppose. But I can't really put my finger on what's what. I had this teacher who thought I should apply for conductor training. But I checked up what it entailed, what kind of qualifications and experience you needed to apply, and I convinced myself that I didn't have anywhere near enough. And then many years later, perhaps 30 years, I found out that the knowledge and skills I had would've sufficed. So you could say that I deceived myself because I didn't think I was good enough.

Maybe Joel wouldn't have been a good conductor, maybe the world would have lost a violinist without getting anything in return. But maybe he'd have conducted one of the world's great orchestras. We'll never know. His fear of failure was too great and the support behind him too weak.

Like Milton Glaser, Joel suggests that the fear of failure is, above all, the fear of having to acknowledge one's own shortcomings:

> The older I get, the worse it feels to be revealed as having flaws. [...] I'm afraid of exposing my own flaws myself, not of someone else doing it. It'd be terrible for me to hear a loud chord – if I'm standing there conducting an orchestra – and not hear, for instance, that it's the fifth in a big chord that's a bit [...] I'll hear that it's flat, but I might not be able to identify it as a fifth. And that's something I don't actually need in my skill set but it's enough that I reveal myself to myself. [...] It's small things like that that have been my mental blocks stopping me progressing.

Even for a person who is artistically talented and professionally adept, like Joel, the risk of seeing your own shortcomings can feel too frightening, so much so that you balk at certain opportunities. And this is despite the fact that he never seriously failed in whatever he undertook to do: 'Whenever I've risked it, I've never had to regret it.' Maybe this is linked with the fact that the keen-sighted more easily spots his or her own flaws and shortcomings than someone who is less acute of eye and mind. One's own talent can erect a kind of hurdle in your path, says Joel, one that no one else can even see.

But there is another aspect of fear that also flashes past in Joel: the fear of *succeeding*. Because success often brings other things that we might not be prepared to put up with. We take a step out of the protective collective, be it an orchestra or any other group of which we have recently been a member. We find ourselves in the limelight, having to sustain a role that we might find uncomfortable, if not unbearable. We are placed outside and above the others whether we like it or not.

The researcher who succeeds will soon find herself proclaiming her message in front of a conference audience despite possibly feeling like a forgotten newcomer. She gets invited to meetings that she feels she doesn't belong in, situations in which everyone else appears more intelligent and knowledgeable than she is. She is expected to appear on scientific panels and participate in debates and to be questioned and criticised for who she is and what she says. By no means everyone is able to subject themselves to all these horrors and they opt, consciously or not, to fail because the companions of success are so unpleasant.

Fear as an engine, fear as a brake. It's so hard to use our fear properly, isn't it? Maybe we should just be happy that we can sometimes put it to productive use rather than just constantly wish to be less afraid. Just look the monster of fear squarely in the eye. If I can take the monster by the hand, despite my fear, maybe it will lead me right, all the way to the brink.

Boredom Is the Mother of Creation

It's my last day in Umeå, at least my last day as professor of sociology. It's been a fantastic day. I've held a valedictory lecture on how one can try to retain a creative disposition throughout one's career. A great response, a great discussion. Then the official thank-you, the people approaching me before going home to wish me luck and say that I meant something to them. I should quit more often, I think to myself, you never get to hear this stuff on a normal day.

I'm now sitting alone in a pub just off campus with a beer in my hand waiting for the taxi to take me to the airport. The dean of the faculty walks in.

He is formally speaking my immediate superior but we've not seen much of each other over the years. 'Hi,' he says with a smile and a nod and sits down at the almost full table next to mine. 'Hi,' I say. That's the sum total of our conversation. Instead, he starts to chat to the people beside him about, as far as I can make out, local university gossip. There's a new rector due in, but was his appointment that kosher? What has he or she said about this or that? My interest in minimal, I have no urge to join in the conversation, and no attempt is made to include me.

It feels a little strange. I'm a researcher at his faculty, I've been working there for 31 years, these are my last moments at the place. Maybe he doesn't know that this very day is my last one, although he's fully aware that I'm on my way out; he's even signed off my letter of resignation. But he prefers the local tittle-tattle. I stand up and start to put my coat on. 'Bye, then,' I smile. 'Bye.' We'll probably never see each other again. I walk out into the slushy snow to await my imminent taxi. The chilly wind picks up, I shiver and feel happy to be on my way.

I left Umeå for the sake of love, but had no regrets doing so. Boredom had settled into both my professional and private existence. I had a good life, but it was mundane and I was stuck in a rut. I knew my research field like the back of my hand, and my colleagues. I often knew what someone around the seminar table was going to say as soon as they raised their hand, and I was rarely surprised by my research. The intellectual environment that had buoyed me up for so long, that had both shaped and been shaped by me had started to feel … maybe not like a straightjacket, more like an item of clothing that was a little too tight.

I was living alone and my pedantry made sure that not a single thing was out of place in my large flat. My ability to plan ahead was such that events had already been foreseen many times before they occurred. Boredom lay like a grey woolly blanket over my well-ordered existence. It was time to pack up and leave.

Boredom kills. But boredom is also a blessing. It's the door to another life.

Creativity and boredom stand in a paradoxical relation to one another. They are often thought of as mutual opposites; the creative person is exciting, fascinating, innovative, perhaps even funny. Not sad and bored or boring.

But the relationship between boredom and creativity is more complex than this. In actual fact, creativity is often born out of boredom, out of simply being unable to bear doing the same thing again in the same way. You start to improvise, push the boundaries, anything to create a little variation. And then all of a sudden something new appears.

In his book *Outliers* Malcolm Gladwell writes about the Beatles' time in Hamburg before anyone had heard of them.[17] They played an unbelievable amount of music at the club that engaged them eight hours a day, seven days a week, week after week. It's a mystery how they managed it. It wouldn't have been possible without the constant improvisation, and it was in these improvisations that what was to become the Beatles' sound emerged. When they arrived in Hamburg: a few lads from Liverpool. When they left Hamburg: a band that would define the future of pop music.

But the story of the Beatles in Hamburg isn't just a song of praise to improvisation. It's also a tribute to endurance. Here too, boredom and creativity are interwoven. To really achieve something novel, one must slog away. To try, try, try again. And again. And again. To not give up, to stick with it. Mastery takes endless repetition. Whoever cannot put up with creativity's inevitable tedium will become trapped in an even greater tedium, one that leads to stagnation and, at worst, spiritual death. More on this horrific state of being later.

Joel Sundin makes a point to stress how important childhood is to the ability to grow as a creative person. It's about being able to live freely, without a set of rules governing what you can and can't do. But it's also about the significant role played by boredom and apathy:

> Not being fed with things the whole time. We need ennui so that things happen in our brain, things we produce much later. At least that's how I think it's been for me. [...] Long periods of ennui and loneliness, that have done me good, I think. These days it's something to be scared of, we're scared that the kids will be lonely, we're scared of them getting bored. And I think there's a danger in that. [...] Boredom is the absolute antithesis of creation and antitheses are vital. I think that to experience real happiness you need an antithesis. And to get the urge to *do* something.

Boredom can therefore be a *precondition* of creativity, but it can also be an *obstacle* to it. Maybe we can think of this through the development of the Aristotelian theory of virtue proffered by the German philosopher of religion Peter Knauer.[18] Aristotle's original idea was that for virtue to come into being, each extremity needs to be balanced by its opposite. To illustrate: between the two polar vices of 'recklessness' and 'cowardice' lies the virtue of 'courage'. But Knauer points out that this is not a balanced triad: 'courage' lies much closer to 'recklessness' than it does to 'cowardice'. Between 'courage' and 'cowardice' lies something else: 'caution'. Every virtue should be balanced by its opposite to prevent it flipping. Courage without caution is recklessness; caution without courage is cowardice.

In our case, we could say that innovation without endurance is fickleness, while endurance without innovation is monotony. The fickle one jumps from one thing to another, leaving nothing of any permanent value behind; but the one who never gets bored becomes stuck in a monotonous rut, never feeling any desire for self-renewal, self-replicating for all eternity.

Maybe all this is about the difference between fleeing and seeking, between giving up or throwing in the towel and wanting something different or fighting on. The one who is bored and powerless flees only to end up back in the same miserable situation; the one who is bored but self-determining can proceed in another, more genuine way.

There is also a research policy point to score here. In recent times, the systems for evaluating what is good research and assessing who among the fiercely competing applicants will get their research financed has taken on a more formalistic aspect. Instead of knowledgeable people reading and evaluating what one has done, all the more attention is given to counting one's achievements: the number of publications, the number of citations, the *impact factor* of the journals in which one has been published. In this trend, the researchers themselves are far from guiltless. We are only too happy to bask in the glow of our prestigious publications, to forever count how many times we've been quoted or referred to. I do this rather a lot and I'm not the only one. My good friend Bo Rothstein's son suggested that on his gravestone, Bo should have a counter that continues to tot up his citations even when he's dead and buried. He was joking, yet wasn't.

Formalistic valuation systems breed egocentrism, cynicism and despondency as well as sophisticated attempts to dodge the system. Publications are broken into fragments to create a larger catchment area for the automated assessors. People with money or control of the data shoehorn themselves in as co-authors despite having contributed nothing to a paper's genesis. Clever *self-promotions* and *search-engine optimisations* are becoming increasingly common and increasingly ingenious in this the era of the web. An attention-creating contest in which, in the long run, we are all the system's losers.

But perhaps the biggest problem is that short-term and formalistic assessment systems foment self-replication. After all, it's nearly always quicker and easier to do more of something we are already good at, things that are the same as something we've already done in all but name. Yet another publication can be cited and counted, yet still boredom comes creeping as a sense of meaninglessness. What's the point, who will actually look at the counter on my gravestone?

For boredom to be the mother of creation, something else is needed. It takes daring to meet the other's gaze, to give and receive help, to leave the path too well-trodden. It takes courage. The environment one works in requires trust and sympathy – the prerequisites for someone daring to grow. This does not mean an uncritical embrace – I might be on the wrong track and it's good if someone tells me so before I've wandered too far into the wilderness. But it does mean that my inevitable insecurity must not be exploited to flatten or crush me. This should go without saying, but I've seen so many research environments acting as the pulverisers of new ideas that I know it cannot be taken for granted.

If the boredom cannot be defeated, either by finding meaning in replicating things to ascertain their veracity or by creating something genuinely new because you can no longer put up with the old, all that awaits is death. Or at least the spiritual death that comes with loss of meaning.

Meaninglessness is constantly lurking in the creative industry. In many other industries, it's easy to see why things need doing, where the immediate utility lies. If my office never gets cleaned I'll find it impossible to spend any time there; if the brakes on my car fail I'll have to get them fixed to avoid crashing; if I don't get food I'll starve. The meaning that inheres to cleaning, repairing cars or farming is thus obvious and immediate.

This is not the case with texts and art. To be sure, we could claim on good grounds that a society without literature, painting, sculpture and music would be an impoverished society or that the texts that researchers produce can create reality in the form of new phenomena or ways of viewing the world. But this meaning is dissociative and derivative. If the texts are to change the world, someone has to read them, assimilate them and put their ideas into practice. And beauty, like creativity, is in the eye of the beholder. If no one ever thinks that the text is worth reading, it isn't readworthy, whatever its author might think.

It is therefore easy in the immediate daily experience of writers and artists for meaninglessness to strike. If it isn't successfully defeated it can cause one to give up and flee the meaninglessness instead – into other activities in which meaning is more obvious or out into abuse of the self or others.

Some researchers – like me – try to convince ourselves that it doesn't matter that much how many people actually read what we write. That we write for ourselves or for the initiated and for posterity, while the approval of the masses is of no value to those of us who don't need to make a living from writing. I maintain that we are hypocrites. What we do acquires meaning only in the heads of others, and if the others aren't there, it can all too easily become meaningless.

Maybe this longing to mean something for someone else is also one of the causes of the frantic citation-totting, however mechanical and almost perverse it can seem. I'm cited, therefore I am. As a means of creating and maintaining meaning in what the researcher is doing, however, I doubt it has much staying power.

Where the meaninglessness arises is from the absolute zero of creativity. It is a terrible place. Life is drained of colour, as if the light has been muted by a hazy greyscale as in those films that tried to depict the misery and ugliness of Eastern European Communist dictatorships. What just a while ago seemed enterprises of great pith and moment are exposed as mere trifles. There's no point in anything, since the very foundations of one's activity have crumbled. Boredom is too weak a word for it; it's a much deeper form of existential crisis than that. I haven't been there often, or for a long time, but I know the place only too well. And I don't wish to go back.

I can't really remember how I escaped. I know I needed a lot of help, but I'm not sure who gave it to me or how. But suddenly colour returned to life, what until recently was without meaning regained its purpose. I could mobilize other emotions, such as anger and fear, to combat the meaninglessness. Maybe this is the deepest signification of creativity. Being able to pull yourself up by your hair and save yourself from spiritual death. To create something new, because the option of staying put is no longer viable.

The Quiet Fury

I'd always found M quite difficult. He had this frequent irritating grin on his face, he wasn't that smart and was useless at sports. He compensated for his weak status by harassing the few people even lower than him on the ladder. I often had the feeling that he would have picked on me too, if he'd only had the courage. But even if I wasn't that high up the totem pole, I was well integrated into both power groups in the class, and this fact, combined with my high grades and success on the field, deterred him from challenging me.

The football match surges back and forth. I get the ball on the left wing, drive a through ball to the right as I sprint towards the opposing goal, get the ball back on a one-two, accelerate and suddenly find space. But I get no further: someone has grabbed my shirt from behind and stopped my advance. I turn around angrily and meet M's grin. Something inside me goes black. I turn and kick him with full force in the hip. He falls and lies on the ground whimpering with tears in his eyes. Ridiculous. I'm immediately sent off, full of righteous indignation, and plonk myself down on the reserves' bench. I'll kick him again if he comes close. I am remorseless.

There are many reasons for a researcher to get angry. There is little room for manoeuvring when it comes to financing your research and publishing it, and you're in a state of constant competition from start to finish. You're forever being judged and not always in a way that you find informed and fair. A research career is replete with fouls.

Researchers are often a little egocentric. And I'm no exception. We live so much in our own heads with our own thoughts and ideas that the lack of understanding shown by those around us can seem almost inconceivable. And a reason for anger.

But rather an angry researcher than a prostrate one.

On a supervisors' course that we both attended, a colleague at Umeå University talked of the difference between her and her male supervisor when they had their scientific papers criticised and refused. The supervisor would hit the roof and pace the corridors spewing bile over all the idiot reviewers and their imbecilic arguments. Then he'd sit down and rewrite.

My colleague, on the other hand, would enter a state of quasi-depression: 'I knew it, I'm not good enough for this job and now they've rumbled me.' All her aggression would be turned inwards rather than outwards. Valerie Young's *impostor syndrome* in all its glory. But then she too would sit down and rewrite – and as far as I know she's still researching.

I hope she's angrier these days.

Research is a form of institutionalised failure. So much of it is dedicated to finding flaws in your own and – above all – others' work. The opportunities to get angry at the inability of your peers to accept what you've achieved are legion.

It can't always be much fun to be an editor when frustrated authors explode over what they see as incomprehensible refusals. We can get a little insight into the tormented world of the journal editors by reading the publication advice given by *Governance* (which for some baffling reason has refused both the articles I've tried to get published in the journal …). One piece of advice is about taking refusal on the chin:

> Handle rejection elegantly. Editors and reviewers are well-intentioned scholars who have taken time to consider your work. If your manuscript is not accepted for review or publication, don't take it as a personal affront. Don't respond immediately to an email notifying you about an unfavorable decision. Take a day before you reply. Never question the good faith or competence of a reviewer. You can ask the editors to elaborate on their reasons for rejecting your submission. But also ask yourself why well-intentioned readers could not see the power of your argument.[19]

What are they saying? We are people too, we do our best, we don't mean any harm, don't be horrible to us. We don't deserve it. One can only imagine the torrents of aggression that gave rise to this well-phrased piece of publication advice.

I myself rarely enter the fray with anonymous reviewers who don't like what I've done. Not even when I think that their assessment is deeply unfair. I know that the chances of winning the battle are infinitesimal and that the reviewer either did their best and will be angered or upset at my displeasure, or is a sadist who's now getting a kick out of watching their victim squirm. I bunch my fist in my pocket but I'm not happy. Moreover, the world of reviewers in a particular field is so small that I can often sense who of the four or five people it is hiding behind anonymity. The terrible thing is that they have all gone down in my estimation, despite only one of them being guilty. My hostility radiates like ripples on a pond.

In *The Managed Heart: Commercialization of Human Feeling* sociologist Arlie Hochschild gives a fascinating account of how anger can be used to tackle a difficult task.[20] Her book deals with emotional work, about how different professions learn to routinely create and handle their visceral responses as part of their practice. Her prime example is flight attendants who learn not to become angry with their passengers, however nasty and disruptive they might be. And how this routine emotional tinkering easily creates psychological problems when the manipulated emotions stop acting as a signal function when confronted by someone's grief, fury and feelings of personal violation.

But my favourite example from the book is the flight attendants' counterpoint: the employees of companies' debt collection departments who learn to switch off their empathy for the people whose outstanding debts are to be settled and unpaid-for things recovered. Every day they hear heartrending stories of death, divorce and family tragedy that not uncommonly are the reason for the defaulted repayments and mounting debts. But if they're to do their job, they can't let themselves be moved or drawn so deeply in that they end up making the indebted's problems their own. So they learn from day one to work themselves up into a cold wrath, a frame of mind in which the people they are dealing with are diminished, reduced to morally degenerate wasteful slackers. This enables them to stay on task, so that the money comes in and those who overspent get to atone for their mistakes.

One might wonder how these people, who work in a perpetual state of cold rage, adjust to their family and friends after the working day is over. It's not hard to imagine that the constant professional summoning of empathiless-ness is not the best grounds on which to build functional relationships. Unfortunately Hochschild's book leaves us none the wiser on this count.

To advocate cold rage as an appropriate frame of mind for research would perhaps be to take things a little too far; but quiet cool anger can actually be a good companion. It helps us to dare a little more, it prevents us from becoming too personal in the construction and delivery of our arguments. If managed properly, anger can make us burn off the 'diplomatic fluff' that Swedish mathematician Olle Häggström wishes to avoid for the sake of clarity in his reasoning.[21]

There are many ways in which anger is a useful resource. The fear of stepping up on stage, which I identified as a reason why some researchers choose not to succeed, can be overcome by anger at the state of things. 'If I don't do this, the baddies will get away with it. I am the only one who can stop them' is a thought that, if you're lucky, can conquer the fear of stepping into a role you feel unable to assume. Anger at the regime of lies, ignorance or vile self-interest can eventually make us do what we dread: step up, assert ourselves, speak up. Claim to have something to say and try to get others to listen.

Anger can also prevent us social scientists from overly appropriating the perspectives held by the objects of our study. *It is difficult to get a man to understand something, when his salary depends upon his not understanding it*, observed author Upton Sinclair caustically about the obliging spirits in the pay of the US corporations.[22] And there sure are enough people whose income is based on their ability to silence their conscience and obey their master's voice. No social scientist worth the title can 'interact' with such people – despite the fact that nowadays we are constantly urged to 'interact' with those we intend to study.

Where many kinds of 'interaction' require us to repress our anger, smile kindly and try to find common ground, clarity of thought often demands that, like Hochschild's furious debt collectors, we're able to refuse to make the others' perspectives our own. That we stand firm by what we believe in even if it's unpopular, that we refuse to yield to anything but facts. And not to be like Hochschild's flight attendants, who at the end of the day no longer know whom they really love or hate, having so effectively learnt to switch off their real emotions.

Self-contempt and Revanchism

My wife has a summer house on the south part of the island of Gotland. What a lucky man, someone might think, his marriage brought him a wife *and* a summer house on Gotland. And it is indeed indescribably beautiful there, the house and its location are indeed unique, and it is indeed very nice to be there. But things are more complicated than that. The cottage stands both in heaven and in hell. It requires constant maintenance, renovation and a whole lot of daily work. Work that either I'm not very good at or, as is more often the case, have absolutely no idea how to do. Every visit elicits the same emotional cycle, from the joy of arrival at the most beautiful place on Earth, via a quickly escalating aversion towards everything that needs doing and creeping self-contempt at how little competence I have as soon as I move away from the office and my computer screen, to a sense of panic over ending up here again, until finally a kind of reconciliation sets in. Sure I'm useless but it's so lovely here that perhaps it doesn't matter. And anyway, we'll be going home soon.

My wife says that a more serious case of performance-based self-esteem would be hard to find. Not even the summer house is spared its crowbar.

Performance-based self-esteem is a deeply felt sense that one's human value is determined by what one does and achieves. A feeling of incompetence isn't just frustrating, it's an existential threat. The strange thing about this feeling of mine is that it primarily concerns myself. Other people's mistakes and ignorance I can often tolerate, but of my own failings I am totally unforgiving.

The ease with which I endure the shortcomings of others is, in other words, something I am unable to apply to myself. I long saw this as a good quality, that I always expected more of myself than I did of those around me. But now I can see the incredible arrogance in which this is rooted: what's good enough for others isn't good enough for me.

Self-contempt can thus in an apparently paradoxical way derive from arrogance and self-pride. It is not primarily the censure and condescension of others that fuels the self-contempt but the fact that one does not live up to one's own (high) demands. When the achievement one finds acceptable in someone else finds no favour because it is one's own; when one is 'forced to face oneself, stand eye to eye in a kind of witness cross-examination with the uselessness of one's writing, with all the disgust, with the thought of how easily one is beguiled by hubris into having something to say, all the wasted time, the emotions ploughed into something stillborn', as Swedish journalist Kristofer Ahlström writes in an interview article about authors and their struggles with self-contempt: 'All the author sees are the flaws, the longueurs, how transparent and fragile the whole construction is. They are sick at the thought of themselves.'[23]

In his book *Rädslan för svaghet* Swedish historian of ideas Sverker Sörlin describes the difficulty he has accepting the weakness he so easily tolerates, indeed almost cherishes, in others. How reluctantly the insight of his own weakness comes to him:

I am weak.

Actually, this is a word I like. I like the weaker. I like redress that is given to those who are trod upon. I like support for the weak. I like the right to be weak. I have always believed that truth and insight are much closer to weakness than to strength. I have worshipped weakness from afar.

But I have not considered myself weak. I am the one who can help. I can guide. I can see potential. Weakness might be unavoidable but it should be tempered for the sake of freedom and personal opportunities. I can help to temper the weakness. The weak are the others, the ones for whom I feel all this empathy and understanding. And anger, when I find people demeaning weakness. My hatred of the political ideas that embody a formulaic contempt for weakness. My hatred of the politics of power, of muscle. Of almost everything that tends to be associated with man as an animal. For we are not animals. I am not an animal.

My own weakness is a non-thing.

But not any more.[24]

For Sörlin it is the fight against gout, a sickness that is as painful as it is ridiculed, that reveals to him his fear of his own weakness. The well-trained athletic body plays constant wicked tricks on him and leads him out into a quest for the cultural correlates of weakness and disease. Maybe the weakness of others had been a means of feeling strong for him, and not at all grounded in genuine solidarity and sympathy. For me, it is having to deal with my own incompetence that has revealed to me my unpleasant self-intolerance and made me wonder how well-founded my view of shortcomings actually is. And that has made me muse over how tolerance and arrogance are actually related. Demanding of oneself that one must always cope – how arrogant is that?

Self-contempt easily comes creeping in the shadow of rejection. On a private and personal plane it can happen when a loved one leaves you, when the people you thought were your friends avoid you, when the one you desire shuns you. In the world of research, and more generally in that of writing, it is about refusal. Articles that the journal finds too substandard, the book manuscript that the publisher refuses to publish, plans and ideas that no one wants to finance. It is easy then for that voice saying *it's finally dawned on them how useless you are* to whisper in your ear. No matter how many times you've done well, sometimes just one refusal does it if it's unexpected and brutal enough.

Still today, after hundreds of messages from journal editors that have reviewed, refused and occasionally published my carefully wrought creations, the letter from the journal in my inbox triggers a tsunami of emotions. The eye scans quickly for the conclusion – is this a refusal or not, is it a refusal with no second chance or am I still in the game? The latter is the most common outcome, and so the next emotional destination is the verdict of the reviewers.

Have they understood what it's about? Are they cruel or constructive? Are their demands fulfillable? Am I out for the count or have I merely suffered a few light jabs to the head?

It has taken me years, decades even, to learn to sift out the tone of reviews from their content. Some reviewers spice their stated opinions with various bitter or directly condescending comments, but when I see the changes they want to see before the paper can be accepted I notice that there aren't that many, and what there is is easily remedied. Other reviewers can be as friendly and objective as anything, but when I look more closely at what they want it is effectively a complete rewrite.

The first article I ever sent to an international journal was accepted as it was and no changes had to be made. I had no idea how incredibly rare this was, so when my next paper was rejected and showered with unreasonably (to me) severe criticism I withdrew it immediately from the journal in order to try elsewhere. An older colleague shook his head when he saw the reviews and heard my decision: 'They were nothing, you could've easily met these demands,' he said. Over the years I've realised how right he was and how my youthful and slightly perplexing combination of arrogance and weak self-esteem had misled me.

My emotional outbursts have gradually had more to do with how stupid, ignorant and malicious the reviewers are, instead of how thick, useless and bumbling I am. But this nagging feeling of having been rumbled never completely passes. I can always smell it in the air, those toxic vapours of self-contempt. You're stupid, you're incompetent, you've had your day. Time to put your tools away.

I send a book manuscript to a good friend whose literary judgement I trust, trust in fact more than my own. I find out when he calls a few weeks later that he hates it. He tells me that almost everything is bad: the author's voice is unclear, as is the point of the book; I overuse metaphors; I dabble at times in trivialities; and I'm prone to talking down to the readers while at the same time foisting upon them an uncalled-for intimacy. The criticism is devastating and it feels terrible. I make a half-hearted attempt to ascribe hidden motives to him that would explain his merciless slating of my work and spare me painful re-evaluation. But it's no good – I know that he wishes me well and I can tell how much it pains him to not feel the slightest bit enthusiastic about my text. My own enthusiasm for what I've written crumbles and I now see the obvious flaws that he's pointed out; in fact I don't understand how I and other beta readers have managed to avoid them.

I seriously consider giving up everything. If the text really is this bad, maybe this isn't the kind of writing I should be attempting to begin with. What hurts

me most of all is the fact that my style has been subjected to such severe examination. As a researcher, I'm inured to sometimes having my work ripped apart
by stiff-necked reviewers. But I'm not used to being criticised for my writing
style – on the contrary, even the most brutal of my critics admit that at least
the text was well-written. I've even been interviewed (as one of many others)
for a book on successful academic writing.[25] But here I sit now, with a text that
has hopelessly lost its vigour and freshness.

I lose sleep for a few nights as I brood upon what's gone wrong, torturing
myself on the matter of pressing the delete button. That I end up not doing
it feels more like a concession to the bad conscience I have towards those who
have looked at the manuscript and found it interesting and worth reading.
But can I really trust their judgement, or even my own? What if the revised
product can be caught with as many flagrant errors and flaws as my first
attempt, and I can't see it? Should this even be published? Why did I set about
doing something I clearly can't do properly? The manuscript lies there, no
longer enticing, a monument to my own incompetence and failing judgement.
But the self-contempt that would come with giving up feels greater than that
which would follow on from my inability to make the text as good as I want it
to be, so I let the one self-contempt vanquish the other and try again.

Self-contempt is a close ally of revanchism. I would go so far as to say that
revanchism commonly appears on the successful externalisation of self-
contempt, when one places it outside the self in the form of an experience of
the other's condescension. Revanchism has the scent of the underdog about
it. It is the authority, in whatever form it may take, that looks down upon me.
And it is the ones to whom I shall prove myself. *I'll bloody well show them* is the
credo of revanchism, hissed through clenched teeth.

Joel Sundin says with a chuckle that 'bitterness' is a driving force in his
business. I ask him in what way:

> Bitterness embodies [...] to feel bitter you need to feel that you're actu-
> ally better than the one you're bitter at. And the sense of knowing this
> is a self-driving force, right? This knowledge that you're actually good at
> something. And then you're prepared to push that load in front of you,
> so to speak, right? [laughs]

Revanchism is born from the feeling of being unjustly passed over or
overlooked. It has its roots in all kinds of subordinations: as a woman in a
male-dominated society, as a minority of colour, sexuality or ethnicity, as a
working-class person in the salons of the wealthy. But even the white, heterosexual, Swedish-born middle-class boy who is Stefan Svallfors can find fertile

soil for his revanchism. The social remove from the circles in working-class inland environs of Norrbotten where I grew up, to those in which I now spend my working day, although not astronomical, is considerable. Still today I can feel rather lost at a cocktail party where there's cheek-kissing and polite but learned small-talk to be done.[26] At times like this I long for my desk or the small conversational situations that I find really meaningful.

I am often gripped by a not entirely appetising smugness when I find that some of those with the most academically perfect social pirouettes lack real perspicuity or well-grounded research experience. That they just stand there 'talking shite', as some might say.

Then I can get the same feeling I had when in my youth as a competitive skier I'd come across people with overly expensive and rarely used equipment out in the tracks. My skies were worn by hundreds of miles, theirs were shiny and new. Their clothes were of the latest cut, mine were in fashion a couple of years ago. But I could ski, they couldn't. My resting pulse was just under 50 beats a minute, their hearts ticked considerably more quickly. Sometimes I'd lag a few metres behind one of these material skiers, just hanging back to see how much I could push up his (it was always a his) pace before passing him with majestic nonchalance, ideally on the uphill slopes. I'd disappear over the crown without so much as a glance behind me. A delightful divertissement in the drudgery of training.

It is the same today with me and the academic poseur, who with grandstanding pretension parades the blindingly obvious or the simply untrue. I'm not talking about students struggling desperately to grasp the almost ungraspable. Them I want to help, to ease their journey. I'm talking about someone who's strayed undeservedly far by virtue of their habitus, their embodied way of being. Someone who tries to use verbal and non-verbal language to assert themselves. Someone with a distinguished past and traditions or an ambitious climber, it doesn't matter; my gaze rests cold and ruthless on whoever attempts to dupe us. 'I wonder why you don't mention this book, which essentially seems to contradict everything you just said,' I mildly muse. 'Is it because you aren't aware of it, or because you don't think it has anything to add in this context?' There's now no decent comeback and I can see in the poseur's eyes that he knows that I know. That the game's over.

Revanchism can turn repugnant when it morphs into hatred and a desire to annihilate the other. Fascism has it as one of its evil roots. But revanchism can also be an engine powering the most fantastic attempts to cross our own boundaries. A way of spurring us on to do more when the impending failure doesn't seem as unbearable as living forever with another's condescension.

A way of daring to get closer to the limit of our own ability. A way of growing. *I'll bloody well show them.*

Angst Is Neighbours with Euphoria

There's only one thing I fear in life, my friend.
One day the black will swallow the red.[27]

It's Monday morning, and I'm being picked up as usual at five o'clock. We have to be there and changed before the parade at 0700 hours otherwise we won't see freedom next weekend. Leave cancelled, as they say in the military. We hardly say a word, mutter something inaudible by way of greeting. As usual, my friend drives very fast, but it doesn't bother me. Or maybe it does, in a way; I'd rather not arrive at all.

I have trouble describing to others, even to myself, why I hate it. Sleeping in a tent, training hard, shooting machine guns, blowing up bridges, driving huge trucks alone – sounds like a 19-year-old's dream, doesn't it? And yet every morning that incredible distress that settles like a weight on my chest, prickling my arms and face. I feel indifferent towards the possibility of us driving into a ditch or hitting a moose in the first light of day. Maybe it would be a relief. I've heard of infantry soldiers who let themselves be injured by comrades out on exercises just to get to go home. Who held up a hand and asked a friend to break their fingers with a rifle butt. I thought it sounded crazy – surely it can't be that bad? Now I'm not so sure any more.

We arrive. I step out of the car, shaking the morning stiffness from my bones, toss my bag over my shoulder, hand over a few 10-kronor notes for the petrol. 'See you on Friday, if not before.' I turn my gaze towards the sapper regiment's yellow and red barracks and feel like throwing up; my arms are still prickling but no one's interested in how I feel so there's no point making a fuss. We'll be falling in soon.

I really don't mean to glorify my angst. Whoever has stared into the abyss and seen that abyss stare back knows that this is no game. The thought of it comes so easily, that last step that puts an end to all suffering. Or that black, oleaginous sensation in the stomach, when the most mundane of everyday actions feels insurmountable and you don't even want to see the people dearest to you. Or when sleep deserts you and refuses to return: the spiritual darkness and icy silence that 4 a.m. can offer after so many sleepless hours I would not wish on anyone, ever.

And yet. If I never felt the darkest sides of life, its lightest sides would not be as sparkling. The darker the shadow, the brighter the light. There's something bovine about people who have never broken down, whose lives know no precipices or sharp edges.

If it can be made meaningful and bearable, angst can be a powerful driver of creativity. Actor and author Evin Ahmad treats the relationship between acting and angst as a question of how one achieves freedom:

> 'To forget yourself, to lose your own filters, like a child. And it doesn't occur often, but when it does, when I'm *immersed* in the role, that's when I'm capable of anything. I can fly, if I want to. There's no better proof of this than coming off stage and not remembering what you've done.'
>
> *Is it the opposite of anxiety?*
>
> 'I think so. When in a state of severe anxiety you want to have nothing to do with yourself, to not think those thoughts. At the same time, reflection is crucial to my work and my anxiety can be found in my book, in all my characters. It's connected to creation. Even if I really don't mean to romanticise. Severe anxiety is absolutely horrible and for me it can be so physical in the form of actually being sick, of getting palpitations. I can't even get out of bed. At such times it's impossible to be creative.'[28]

From the other side of the stage, director Ingmar Bergman describes how his angst – his 'demons' – somewhat paradoxically propelled and supported his artistry:

> Although I am a neurotic person, my relation to my profession has always been astonishingly non-neurotic. I have always had the ability to attach my demons to my chariot. And they have been forced to make themselves useful. At the same time they have still managed to keep on tormenting and embarrassing my private life. The owner of the flea circus, as you might be aware, has a habit of letting his artists suck his blood.[29]

I think that many others, like me, think along the same lines, even if none of us is a Bergman. That we can let our personal anxieties dampen the torment of writing. Some of my colleagues find it extremely hard to get things finished, being unable to accept that this is all they can do for the moment, that it is this that they will have to defend if need be. I rarely find it hard to add that final full stop. When the manuscript has been submitted to the publisher or the article to the journal, I always have a feeling that I've achieved something really worthwhile. Even though the feeling is only ephemeral and soon clouded by (my own or other people's) objections that expose the text's imperfections – there it is, bright and clear, the moment I press 'send'. For a moment I am truly liberated, and free. And I'm rarely as happy as when I can write, see the characters appear on the screen as if by magic, read sentences I've just written and see that they are good, that I am in my essence.

When I write, I constantly long for this euphoric state of being. At its most intense, the rest of my life can seem an unwelcome distraction in my ongoing relationship to my own text. Now I have to go home, make dinner, do my laundry and spend time with family and friends. May it soon be morning again so that I can pick up where I left off! I realise that at times like this I'm possibly not the ideal family member. It's the text and me, and we're doing just fine.

But outside my textual relationship, the demons keep tormenting and troubling me. I suffer from what the experts call SAD, and every year around the first of November my journey into the darkness begins, a journey that I've been trying to thwart with medication for the past few years. Mine is apparently a fairly mild form of the condition – I daren't even think how a severe form might manifest itself. For a long time, I lived under the delusion that a childhood spent north of the Arctic Circle would make me immune to seasonal problems, that my physiognomy would have adapted to the disappearance and return of the light. But not anymore.

Without this torment of mine, my writing would, I think, have been worse. Knowing how it can be when things are feeling really bad makes the threat of judgement seem less terrifying. The negative emotions that a refusal or malicious review would engender are, in spite of everything, a gentle breeze compared with the black storm of mud and oil that angst can send through my mind. So I am bolder in my writing, especially during my lowest periods. What others might think then carries no weight whatsoever and I am strangely completely free in my misery.

And never have the torments of my soul prevented me from writing. On the contrary, in difficult times my writing has been my only solace, the only place where I haven't felt scared and tormented. I am writing parts of this book in the clutches of such an autumn depression. For the past few weeks, I have walked in a grey fog; it feels as though I'm living in a bell jar, as though my life and I are at an infinite distance from other people. A good friend, a psychologist who suffers from the same disorder, calls this 'feelings of depersonalisation' – which perfectly encapsulates what I feel. It's horrible. Fatigue and misery know no bounds and almost everything seems insurmountable. Even though I know nothing of this is my fault, I have a bad conscious for letting this darkness infect my loved ones.

The only time I feel okay is when I'm writing. Only then am I not afraid, to paraphrase this book's motto. And not only that. My indifference towards the others is so great that the text is the only thing that reaches me. My loyalty towards my arguments and phrasings is total, the future judgements of its readers means nothing, my own relationship with my text everything. My text and I exist in a separate space where we may be left in peace and where the rest of the world is immaterial. A place where I know what to do and where I imagine that I can see more clearly than many others – since I have no delusions about a constantly smiling sun.

Sometimes it frightens me that my angst seems to be gaining ground, that its troughs are getting deeper, its bad periods longer, my recovery harder. Will the black finally swallow the red, plunge me into a state from which no amount of writing can rescue me? Or will it continue to help me dare to do things I'm actually too timid to take on? Angst is a dangerous playmate, and I can never know what it intends to do with me. I can just hope.

I understand from the deeply sympathetic look in the eyes of the 15-year-old, eyes that are usually defiant or a little dulled, how much in a bad way I must seem, standing there helplessly in the kitchen. I can't even cope with deciding what to make for dinner or writing a shopping list. I'm at a total loss, I'm a total wreck. Someone I don't know or can't even define has placed an infinitely long ladder in front of me and told me to climb, even though I have no legs.

But soon I'll be writing again, and all will be well.

Learn to Fail

'And now,' I say, pausing for effect, 'here's a diagram representing an elephant!' I've got their attention, or maybe there are still some in the back row who have mentally checked out. But most of them are indeed captivated, I can see it in their eyes. They laugh at the right places and frown when global child mortality appears. The graph represents changes in income around the world, but funnily enough it looks like an elephant.[30] It shows how conditions for large parts of the world's middle strata have improved over the years, particularly the burgeoning middle classes of India and China. But it also shows how little has changed for the poorest, how the global elite is pulling away and how the middle-income bracket in the rich world has completely stagnated. 'It's fertile soil for political discontent,' I say pointing to the dip in the curve. 'It's from this trough that support for the likes of Trump and Le Pen comes.' They interrupt me with questions, can't contain themselves, want to know. Everything is meaningful. I am happy, in the midst of my working day.

The important question is not whether or how often we are tormented by unpleasant emotions. The most important thing is what we do with these emotions, what they do with us and what we let them do with us.

After many years as a supervisor and research leader, I've started to think that this is perhaps the single most important characteristic of the researcher's personality: how to relate to negative emotions, those that follow the inevitable failure? Because things never turn out how you expect, otherwise it wouldn't really be research. And it's never as good as you expect, either because your abilities fall short or because reality gets in the way and refuses to deliver the results you've dreamed of.

A colleague in a related discipline (who prefers to remain anonymous) told me of how they once took on two individuals as doctoral students, two individuals who were impossible to distinguish between in terms of qualifications and qualities. They'd both written excellent student dissertations – otherwise they would never have been considered. They both had interesting and well-thought-out research plans. They both made an excellent impression when interviewed. They were impossible to rank, but fortunately they could both be offered a place. They are of the same sex, are of the same age and have similar backgrounds. They were even assigned the same supervisor.

A few years later, and they're no longer equal. One of them has moved onwards and upwards and now does things that many of the more senior researchers can't do, has written an excellent thesis and has a career staked out. The other has completely stagnated and after four years is turning out papers that are no better than their student dissertation, in some respects worse. With the wisdom of hindsight, what should my colleague have seen but didn't? When they look back, there's nothing identifiable that can explain how things turned out.

I think that this inability to account for what happened – even with the benefit of hindsight – is because it was not in intellectual ability that they differed – they are both very gifted and good at what they do; it was in the way they handled their minor failures along the way. One of them could exploit failure, turn it around, use it as a springboard. They might not have found what they were looking for along the road but found other things that proved equally useful.

The other one, however, got stuck in failure, which sullied their entire thesis project and trapped them; they brooded, lost heart, thought that everything seemed so pointless. Maybe the thesis project was ultimately too personal so that the results that refused to appear became a kind of existential threat. And time just ticked inexorably away.

This is something, of course, that is almost impossible to predict. Student dissertations carry only faint traces of the author's inner life. All gifted people know how to answer the interview question 'how do you take setbacks and criticism', and can vividly describe how they've successfully managed previous failures. They feel convinced that they'll have no trouble doing the same now too. But not everyone can deliver on their promises.

Whoever has trouble enduring the small failures is therefore heading at full speed towards the great failure; whoever can more easily handle failure avoids this fate. Life can indeed seem so unfair.

Handling failure is often a matter of 'compartmentalising' it, of being able to trap it in time and space so that it doesn't drag the entirety of what you've achieved and who you are into the abyss. This is a fundamental concept of that branch of psychology that deals with 'learned optimism' and 'positive thinking'. Even if there is good reason for scepticism towards the miscreants within this genre, where the individualisation of societal problems and jaunty clichés trip over each other, there is equally good reason to take the more sober variants – such as one of the genre classics, Martin Seligman's *Learned Optimism: How to Change Your Mind and Your Life* [31] – seriously. It would be wrong to imagine that the ability to handle failure is carved in stone from birth or early infancy. More productive approaches to inevitable failure are in fact learnable.

Without ever having read Martin Seligman, Joel Sundin describes diffidently yet astutely the psychological process that kicks in when things don't turn out so well for him:

> Firstly, I say to myself: 'That was a one-off, it doesn't have to be like this next time'. And then I convince myself that there were a series of unfortunate events that caused things not to turn out well, and so there's probably no danger of all these unfortunate events following the same pattern again. I have actually thought about this, because it happens now and then. And so I try to link what's happened with the emotional state I've been put in. And I actually try – and it's interesting you say this because I've not put this into words before – I try to separate my emotion from the parameters that led up to it, I sever the ties binding them. [...] What's happened has led to my emotion, to how I feel, and that connection I try to nip in the bud. And [...] hope for better luck next time [laughs]. [...] Ten years ago I could get panic-stricken when it happened, but I've learnt to just let it dissipate. Because if you've experienced it enough times, you know it will eventually dissipate. It can take a few days, perhaps a week, but before you know it you have a new blank sheet in front of you.

The events that made things turn out as they did were exceptional ones, and the failure is unique to this particular constellation of occurrences. The powerful negative emotions can and should be divorced from what happened, tomorrow is the first day of the rest of my life. There's nothing wrong with me.

The difficulty with the Seligman–Sundin way of handling failure is allowing the negative emotions to tell their edifying story without letting them get the upper hand. Otherwise the same mistake can be repeated again and again, because you've become too successful at not suffering when they occur. Rather than a series of tricks, this is therefore about having got to know yourself, intimately, as a professional animal and a creative person. The way there might not be easy, but nor is it impassable.

The journey can be – and appear – easier if we know that others have trod the same path before us. A brilliant approach is therefore the 'CV of Failures' which pops up here and there on the web. Economist and psychologist Johannes Haushofer explains the concept:

> Most of what I try fails, but these failures are often invisible, while the successes are visible. I have noticed that this sometimes gives others the impression that most things work out for me. As a result, they are more likely to attribute their own failures to themselves, rather than the fact that the world is stochastic, applications are crapshoots, and selection committees and referees have bad days. This CV of Failures is an attempt to balance the record and provide some perspective.[32]

If we were all a little more open with everything we're not successful at, we might not feel such failures with things go wrong. An industry such as a research that by its very nature is one of failure should be able to embrace this idea a little better. Instead, we try to keep a stiff upper lip and suffer in silence. The doctoral students or postdocs who desperately seek an income and a future can be excused any reluctance to wear their failure and pain on their sleeves. But what excuse do the rest of us have?

Handling negative emotions takes finding the *lagom*, the state of not too much, not too little; it is a delicate balance that must be achieved and maintained. Too little fear breeds complacency, too much is paralysing. In just the right dose, it spurs us on to go that extra mile. Anger and revanchism are important drivers as long as they don't reduce down to hatred and a desire to crush and destroy the other. Angst can both be tempered by and encourage assiduous creativity, but in larger doses it is unbearable and will silence us for good.

Finding this *lagom*, this fulcrum of the inquiring life, therefore requires us to find our true selves beyond the smokescreens and sidesteps. But how? I don't know, but I do know that there aren't many people who can do it unaided. Most of us need help. The help can come in the form of a series of conversations discussing the researching life as it is actually lived, or as texts showing that I'm not alone in my condition. Help can come from my trying to help others and in their predicament recognising echoes of my own.

If help is to come, the first step must be that we stop lying, both to ourselves and to others. That we stop talking about research and writing as if they could be done by the book and according to schedule, in passionless sangfroid and blithe confidence. That's not how things work, and as long as we continue to tell these imbecilic stories, we're just helping to make life difficult for ourselves and others.

In writing this book I have no therapeutic or pedagogical intention in mind, other than possibly the purely self-therapeutic. It is not a manual or a guarantee certificate. It is a report from a borderland, where I've been living for so long but where I've only recently been asked to talk about what I've seen and experienced. Maybe it's creeping old age that's made me embroider and remember, as is age's wont.

But what I really believe to be the driving force behind this egocentric narrative is the nasty feeling that we're heading in the wrong direction. That the truly human aspect of research and knowledge production is becoming increasingly silenced or denied. That we researchers are being reduced, by ourselves and others, to incentive-driven research robots. That we are so willingly complicit in the lies and so gratefully receive the superficial affirmation. That we so easily sweep truth and beauty aside as soon as someone waves a carrot or a stick at us.

The simple virtues. To be bold in the face of fear. To do your best. To accept the help that's offered and offer what help you can. To admit to and learn from mistakes. To care about yourself and others. It's no harder than that – or perhaps that hard.

How else will we get there? To the truth, the goodness, the beauty? The place towards every inquiring and writing person should strive. Towards not outward but inward affirmation based on being able to benefit others and to live as one with our true natures, which Aristotle argued was a precondition of the highest form of happiness or *eudaimonia*.[33] Why would we want anything else?

Fear, boredom, anger, angst, self-contempt. The times I've despaired over how hard everything seems to be. And yet. I wouldn't want it any other way. If I could choose again – in full knowledge of what it would be like – my career choices would be pretty much the same as those I've now made. This is what I know and want. *This way, or no way*, sings David Bowie on his last album, and I think I know exactly what he means. Even though this book is often about the dark sides of research, it is nonetheless still the light that I remember most clearly, those moments of serene bliss.

I look up from the screen. A productive day of fresh insights is drawing to a close. Outside twilight is settling in. I'm tired but happy, I feel that it's been a valuable day, that I myself am valuable. That brittle crystal clarity of the late autumn air as I step out onto the street in the darkness that has already fallen, that feeling of limpidity and space. Of meaning and belonging. Tomorrow await new failures, new fears, new angst, boredom and anger. A battle that will never end until it ends. But that's tomorrow, and right now, everything's just fine. Breathe. *Life.*

NOTES

The Cathedral on the Plain

1 Film director Ingmar Bergman's story of Chartres Cathedral first appeared in his essay 'Det att göra film' (1954), which was originally presented as a lecture at Lund University. It was also published in 'Filmskapandets dilemma' in *Hörde ni?* (No. 5, May 1955). The quote used here comes from an acceptance speech in 1965.

2 Torbjörn Tännsjö, *Vänsterdocenten* [The Left-Wing Reader] (Stockholm: Fri tanke förlag, 2016), p. 245. It is uncertain whether Stendhal actually said this, and if so, where.

3 K. Robinson. *The Element: How Finding Your Passion Changes Everything* (London: Penguin Books, 2009), p. 92.

Chapter 1 Alone Together

1 Samuel Becket, *Worstward Ho* (1983) quoted from https://genius.com/Samuel-beckett-worstward-ho-annotated [accessed 16 May 2019].

2 Ken Robinson says this at about 10:20 in his lecture on www.youtube.com/watch?v=iG9CE55wbtY [accessed 16 May 2019].

3 H. L. Dreyfus and S. E. Dreyfus, *Mind over Machine: The Power of Human Intuition and Expertise in the Era of the Computer* (New York: Free Press, 1986).

4 L.-E. Björklund, *Från novis till expert: förtrogenhetskunskap i kognitiv och didaktisk belysning* [From Beginner to Expert: Experience-Based Knowledge in Cognitive and Didactic Perspective]. Linköping, Linköping University, Department of Social and Welfare Studies (2008), p. 97.

5 G. Kasparov, 'The Chess Master and the Computer', *New York Review of Books*, 11 February 2010, pp. 16–19.

6 M. Gladwell, *Blink: The Power of Thinking without Thinking* (New York: Little, Brown, 2005), pp. 3–8.

7 G. Gigerenzer, *Gut Feelings: The Intelligence of the Unconscious* (New York: Viking, 2007); *Rationality for Mortals: How People Cope with Uncertainty* (Oxford: Oxford University Press, 2008).

8 Björklund, *Från novis till expert*, pp. 102–7.

9 Björklund, *Från novis till expert*, p. 101.

10 I no longer recall what TV programme this was said in, but the story is told in another form in Bob Hansson's book *Kärlek – hur fan gör man?* [Love – How the Hell Do You Do It?] (Stockholm: Wahlström & Widstrand, 2008).

11 M. Gladwell, *Outliers: The Story of Success* (New York: Little, Brown, 2008).

12   Björklund, *Från novis till expert*, p. 127, cites S. L. Beilock, , C. A. Kulp, L. E. Holt, and T. H. Carr, 'More on the Fragility of Performance', *Journal of Experimental Psychology*, 133:4 (2004): 584–600.

13   B. Flyvbjerg, 'Sustaining Non-Rationalized Practices', *Praxis International*, 11 (1991): 93–113, p. 94.

14   Björkman, *Från novis till expert*, p. 144.

15   Flyvbjerg, 'Sustaining Non-Rationalized Practices', p. 95; *Making Social Science Matter: Why Social Inquiry Fails and How It Can Succeed Again* (Oxford: Cambridge University Press, 2001), pp. 14–15.

16   E. Stolper, et al. 'The Diagnostic Role of Gut Feelings in General Practice. A Focus Group Study of the Concept and Its Determinants', *BMC Family Practice*, 10 (2009):17; 'Gut Feelings as a Third Track in General Practitioners' Diagnostic Reasoning', *Journal of General Internal Medicine*, 26 (2011): 197–203; 'The "Sense of Alarm" ("Gut Feeling") in Clinical Practice. A Survey among European General Practitioners on Recognition and Expression', *European Journal of General Practice*, 16 (2010): 72–74.

17   E. Falkenström, *Verksamhetschefens etiska kompetens* [The Ethical Competence of the Health Care Manager] (Stockholm: Stockholm University, 2012).

18   P. Gärdenfors, *Den meningssökande människan* [The Meaning-Searching Human] (Stockholm: Natur & Kultur, 2006), p. 51.

19   S. Sörlin, *Kroppens geni: Marit, Petter och skidåkning som lidelse* [The Genius of the Body: Marit, Petter and Skiing as Passion] (Stockholm: Weyler Förlag, 2010).

20   Per Olov Enquist, *Nedstörtad ängel* (Stockholm: Norstedts), p. 29. Extract from *Downfall – A Love Story* translated by Anna Paterson.

21   E. Falkenström, *Att vara fri och höra till: om viljan att vara sig själv och viljan till anpassning* [Belonging and Being Free: On the Will to Be Yourself and the Will to Adjust] (Stockholm: Telegram Bokförlag, 2004), pp. 77–78.

22   I. Josefson, *Kunskapens former: det reflekterade yrkeskunnandet* [The Forms of Knowledge: Reflected Occupational Skills] (Stockholm: Carlsson, 1991). *Läkarens yrkeskunnande* [The Occupational Skills of Medical Doctors] (Stockholm: Studentlitteratur, 1998).

23   Sörlin, *Kroppens geni*, p. 321.

24   T. M. Amabile, *Creativity in Context: Update to 'the Social Psychology of Creativity'* (Oxford: Westview, 1996), chapter 7.

25   F. Castles, 'The Political Sociology of the Welfare State', *Acta Sociologica*, 51 (2008): 75–76.

26   Rogers Hollingsworth told the story at a seminar in Umeå on 23 November 2001.

27   K. Robinson, *The Element: How Finding Your Passion Changes Everything* (London: Penguin Books, 2009), pp. 117–23.

28   R. Hollingsworth, E. J. Hollingsworth and D. M. Gear (in preparation). *Excellence and Creativity in American Research Universities*.

29   L. Bennich-Björkman, *Organising Innovative Research: The Inner Life of University Departments* (Oxford: Pergamon Press, 1997).

30   Bennich-Björkman, *Organising Innovative Research*, p. 3.

31   Christer Sandahl's figures are from a presentation at Lund University on 'Strategic Research Fields', 111214.

32   G. Gigerenzer, *Rationality for Mortals: How People Cope with Uncertainty* (Oxford: Oxford University Press, 2008), p. vi.

33   The building is described at www.archdaily.com/99198/hedley-bull-centre-lyons/ [accessed 14 May 2019].

34   M. Gladwell, *Blink: The Power of Thinking without Thinking* (New York: Little, Brown, 2005), p. 117.
35   J. Elster, *Sour Grapes. Studies in the Subversion of Rationality* (Cambridge: Cambridge University Press, 1983), part II.
36   L. Igra, *Den tunna hinnan: mellan omsorg och grymhet* [The Thin Membran: Between Care and Cruelty] (Stockholm: Natur & Kultur, 2003).
37   K. Widerberg, *Kunskapens kön: minnen, reflektioner och teori* [The Gender of Knowledge: Memories, Reflections and Theory] (Stockholm: Norstedts, 1995).
38   This is one of the main themes in e.g. D. Josefsson, and E. Linge, *Hemligheten: från ögonkast till varaktig relation* [The Secret: From Glance to Lasting Relation] (Stockholm: Natur & Kultur, 2008).
39   Falkenström, *Att vara fri och höra till*, p. 162.
40   Dalton Conley, *Elsewhere, U.S.A.: How We Got from the Company Man, Family Dinners, and the Affluent Society to the Home Office, BlackBerry Moms, and Economic Anxiety* (New York: Pantheon Books, 2009).
41   Hans L. Zetterberg, 'Scientific Acedia', *Sociological Focus*, 1 (1967): 139–51.
42   Zetterberg, 'Scientific Acedia', 139–51.
43   P. O. Enquist, *Ett annat liv* [Another Life] (Stockholm: Norstedts, 2008), p. 335.
44   L. Bennich-Björkman, *Organising Innovative Research*.
45   Amabile, *Creativity in Context*.
46   The head of research wishes to remain anonymous.
47   A fascinating study of al-Qaida and related organisations can be found in S. Atran, *Talking to the Enemy: Violent Extremism, Sacred Values, and What It Means to Be Human* (London: Allen Lane, 2010).
48   Flyvbjerg, 'Sustaining Non-Rationalized Practices', pp. 93–113. The quotes can be found on pp. 98 and 96, respectively.
49   J. Habermas, *Theorie des kommunikativen Handelns* (Frankfurt am Main: Suhrkamp, 1981).

Chapter 2 In the Corner

1   From the movie *The Matrix*, 1999.
2   Dalton Conley, *Honky* (New York: Vintage, 2000).
3   Chad Alan Goldberg, 'Robert Park's Marginal Man: The Career of a Concept in American Sociology', *Laboratorium*, 4:2 (2012): 199–217. http://soclabo.org/index.php/laboratorium/article/view/4 [accessed 16 June 2014].
4   Helen Mayer Hacker, 'Women as a Minority Group', *Social Forces*, 30:1 (1951): 60–69. (citerad från Goldberg 2012 ovan).
5   Friedrich Nietzsche, *Så talade Zarathustra* (Umeå: H:ström Text Kultur, 2009) (German original, Also Sprach Zarathustra, 1883).
6   The exchange about Gladwell's family appears about 5 minutes into the interview at http://www.theguardian.com/commentisfree/video/2013/oct/24/stephen-fry-malcolm-gladwell-video-interview?CMP=twt_gu [accessed 15 May 2015].
7   'My Story Isn't about Auschwitz, It's about Life after Auschwitz', http://www.haaretz.com/life/arts-leisure/.premium-1.574087 [accessed 15 May 2015].
8   Eric Hobsbawm, 'Enlightenment and Achievement: The Emancipation of Jewish Talent since 1800', in *Fractured Times. Culture and Society in the Twentieth Century* (London: Little, Brown, 2013).

9   Richard J. Herrnstein and Charles Murray, *The Bell Curve: Intelligence and Class Structure in American Life* (New York: Free Press, 1994). For a compelling critique of Herrnstein and Murray, see Claude S. Fischer, Michael Hout, Martín Sánchez Jankowski, Samuel R. Lucas, Ann Swidler and Kim Voss, *Inequality by Design: Cracking the Bell Curve Myth* (Princeton, NJ: Princeton University Press, 1996).

10   Tony Judt, 'Edge People', *New York Review of Books*, 25 March 2010.

11   Gunilla Gerland, *En riktig människa* (Stockholm: Cura, 1996), pp. 161, 245. Extract from *A Real Person* translated by Joan Tate.

12   Gerland, *En riktig människa*, p. 211.

13   Gerland, *En riktig människa*, p. 248.

14   Ronald S. Burt, 'The Social Capital of Structural Holes', in *The New Economic Sociology: Developments in an Emerging Field*, (ed.) Mauro F. Guillèn (New York: Russell Sage Foundation, 2002), pp. 148–92.

15   Roland Perry, *Monash: The Outsider Who Won the War* (Sydney: Random House, 2004). Thanks to Bo Rothstein, who drew my attention to this story and its significance for my book.

16   Nancy C. Andreasen 'Secrets of the Creative Brain', *The Atlantic*, July/August, 2014. http://www.theatlantic.com/features/archive/2014/06/secrets-of-the-creative-brain/372299/.

17   Roopa Unnikrishnan, 'Malcolm Gladwell on the Key to Success: Don't Be Afraid to Look Like a Fool', *Quartz*, 23 October 2014. https://qz.com/283945/malcolm-gladwell-on-the-key-to-success-dont-be-afraid-to-look-like-a-fool/ [accessed 1 November 2014].

18   'Please Call Me, Baby' from the album *The Heart of Saturday Night*, 1974.

19   'The Warmest Room' from the album *Talking with the Taxman about Poetry*, 1986.

20   Rudyard Kipling, *The English Flag*, 1891. http://www.kiplingsociety.co.uk/poems_englishflag.htm [accessed 18 June 2014].

21   See his lectures from Ljubljana at http://videolectures.net/risc08_ljubljana/.

22   Neil McLaughlin, 'Optimal Marginality: Innovation and Orthodoxy in Fromm's Revision of Psychoanalysis', *The Sociological Quarterly*, 42 (2001): 271–88.

23   The beginning of Robin Cook's speech in the House of Commons on 17 March 2003 on his retirement as foreign minister can be seen at www.youtube.com/watch?v=I0f8NBlmwwE [accessed 15 May 2015].

24   C. H. Hermansson, *Monopol och storfinans: de 15 familjerna* [Monopoly and High Finance: The 15 Families] (Stockholm: Rabén & Sjögren, 1965).

25   Sverker Gustavsson, 'Akademisk liberalism', *Statsvetenskaplig tidskrift*, 112:4 (2010): 429–39.

26   This observation was made to me by historian Henrik Berggren in a lunchtime conversation that was actually about something completely different. Neither of us could find a reliable source for the argument at the time, but long after I came across it in Howard Gardner's *Creating Minds: An Anatomy of Creativity Seen through the Lives of Freud, Einstein, Picasso, Stravinsky, Eliot, Graham and Gandhi* (New York. Basic Books, 2011 [1993]).

27   See Gardner, *Creating Minds*, chapter 5, and the article 'Nightmare at the Picasso Museum', *The Guardian Weekly*, 16 October 2014. http://www.theguardian.com/artanddesign/2014/oct/16/-sp-nightmare-picasso-museum-paris [accessed 31 Oct 2014].

28   Per Olov Enqvist, *Livläkarens besök* (Stockholm: Bonniers, 1999). I deal with Enqvist's analysis of power in this novel in my chapter 'Maktens mekanik – reflektioner kring Per Olov Enqvists *Livläkarens besök*' in *Sociologi genom litteratur*, ed. Christofer Edling and Jens Rydgren (Lund: Arkiv, 2015).

29   Gardner, *Creating Minds*, p. xvi.

Chapter 3 Darkness and Light

1   From *Star Wars: The Empire Strikes Back*, 1980.

2   A blistering criticism of the genre is Jenny Damberg, 'Det ska vara en förlorare i år', *Arena*6 (2014).

3   A useful overview of the sociology of the emotions is Jonathan H. Turner, 'The Sociology of Emotions: Basic Theoretical Arguments', *Emotion Review*, 1:4 (2009): 340–54. A comprehensive standard textbook is Jonathan H. Turner and Jan E. Stets, *The Sociology of Emotions* (Cambridge: Cambridge University Press, 2005).

4   See the survey in one of the few books I have found that make the emotions of the researcher the subject of analysis, Caroline Clarke, Mike Broussine and Linda Watts, *Researching with Feeling. The Emotional Aspects of Social and Organizational Research* (Milton Park: Routledge, 2015), pp. 1–11. Their book's main concern is understanding how the researcher's emotions affect or even shape the actual research process, in particular the interpretation of its results. My argument is, instead, about the significance of the negative emotions for creating, undermining or maintaining meaning in our activities.

5   Jack Barbalet, 'Science and Emotions', *Sociological Review*, 50, no. S2 (2002): 132–50.

6   For example, in the autobiographical texts cited in Jack Barbalet's article (note 92).

7   Henrik Berggren, 'Jan Björklunds forskningspolitik är humbug', *Dagens nyheter* 120905, http://www.dn.se/nyheter/henrik-berggren-jan-bjorklunds-forskningspolitik-ar-humbug/ [accessed 10 Oct 2013].

8   Joseph Conrad, *Heart of Darkness*. The transformation of the ivory trader Kurtz into the American Colonel Kurtz (played by Marlon Brando) was the doing of Francis Ford Coppola in his Vietnam epic *Apocalypse Now* (1979).

9   Oliver Burkeman, 'This Column Will Change Your Life: Precrastination', *Guardian Weekly*, 140705, http://www.theguardian.com/lifeandstyle/2014/jul/05/this-column-will-change-your-life-precrastination [accessed 10 July 2014].

10 David A. Rosenbaum, Lanyun Gong and Cory Adam Potts, 'Pre-Crastination. Hastening Subgoal Completion at the Expense of Extra Physical Effort', *Psychological Science*, 25:7 (July 2014): 1487–96.

11 Dalton Conley, *Elsewhere, U.S.A: How We Got from the Company Man, Family Dinners, and the Affluent Society to the Home Office, BlackBerry Moms, and Economic Anxiety* (New York: Pantheon Books, 2009).

12 Valerie Young, *The Secret Thoughts of Successful Women: Why Capable People Suffer from the Impostor Syndrome and How to Thrive in Spite of It* (New York: Crown Business, 2011). For a nice discussion in Swedish on Young's book from a PhD student's perspective, listen to the first episode of Moa Eriksson and Sara Kalucza's 'Doktorandpodden', http://doktorandpodden.se/?podcast=avsnitt-1-impostor-syndrome   [accessed   10 Oct 2017].

13 Roopa Unnikrishnan, 'Malcolm Gladwell on the Key to Success: Don't Be Afraid to Look Like a Fool', *Quartz*, 23 October 2014, https://qz.com/283945/malcolm-gladwell-on-the-key-to-success-dont-be-afraid-to-look-like-a-fool/ [accessed 14 May 2016].

14 Louise Montello, 'The Perils of Perfectionism', *Allegro*, Volume XCIX, No. 8 September, 1999. http://www.local802afm.org/1999/09/the-perils-of-perfectionism/ [accessed 2 Feb 2017].

15 Milton Glaser – 'On the Fear of Failure'. https://vimeo.com/23285699 [accessed 10 Sep 2016].

16 'Stellan Skarsgård: För att bli bra måste du vara ömtålig', Dagens nyheter, 20160102, http://www.dn.se/kultur-noje/stellan-skarsgard-for-att-bli-bra-maste-du-vara-omtalig/ [accessed 15 Jan 2016].

17 Malcolm Gladwell, *Outliers: The Story of Success* (New York: Penguin, 2009), pp. 53–56.

18 My take on Peter Knauer's ideas is based on Erica Falkenström, *Vårdchefens etiska dilemman. Strategier för en bättre praktik* [The Ethical Dilemmas of Health Care Managers: Strategies for a Better Practice] (Stockholm: Natur och Kultur, 2016), pp. 185–86. Knauer's original can be found in *Handlungsnetze. Über das Grundprinzip der Ethik* (Frankfurt am Main: Books on Demand, 2002), pp. 24–27.

19 This text could be found in *Governance's* 'Advice to Authors', but unfortunately has been removed:http://onlinelibrary.wiley.com/journal/10.1111/%28ISSN%291468-0491/homepage/ForAuthors.html [accessed 10 March 2016].

20 Arlie Russell Hochschild, *The Managed Heart: Commercialization of Human Feeling* (Berkeley: University of California Press, 2012 [1983]).

21 See, e.g., http://haggstrom.blogspot.se/2014/03/om-nagra-av-reaktionerna-pa.html, note 5 [accessed 10 Sep 2015].

22 Upton Sinclair, *I, Candidate for Governor: And How I Got Licked* (Berkeley: University of California Press, 1994 [1935]), p. 109.

23 Kristofer Ahlström, 'Författare som hatar sig själva', Dagens nyheter, 161119 [CE 19 Nov 2016], p. B8-9, http://www.dn.se/dnbok/forfattare-som-hatar-sig-sjalva/ (accessed 25 Nov 2016].

24 Sverker Sörlin, *Rädslan för svaghet. En berättelse om sjukdom, smärta och löje* [The Fear of Weakness: A Story of Illness, Pain, and Ridicule] (Stockholm: Weyler, 2014), pp. 22–23.

25 Helen Sword, *Air & Light & Time & Space: How Successful Academics Write* (Cambridge, MA: Harvard University Press, 2017).

26 A compelling cultural-critical account of a similar social journey to mine can be found in Lilian Rydh's *Vi åt aldrig lunch* [We Never Had Lunch] (Stockholm: Dialogos förlag, 2004).

27 Alfred Molina as artist Mark Rothko in John Logan's play 'Red', premiered at Donmar Warehouse, 8 December 2009.

28 Matilda Gustavsson: 'Evin Ahmad: 'Jag hittade den där platsen som alla letar efter'', *Dagens nyheter*, 23 September 2017, http://www.dn.se/kultur-noje/evin-ahmad-jag-hittade-den-dar-platsen-som-alla-letar-efter/ [accessed 30 Sep 2017].

29 Ingmar Bergman, *Bilder* [Images] Norstedts förlag, Stockholm, 1990. Quote from http:/www.ingmarbergman.se/verk/persona [accessed 10 Oct 2017].

30 Branko Milanovic, *Global Inequality: A New Approach for the Age of Globalization* (Cambridge, MA: Harvard University Press, 2015), p. 11.

31 Martin E. P. Seligman, *Learned Optimism: How to Change Your Mind and Your Life* (third edition) (New York: Vintage Books, 2006 [1990]).

32 http://www.princeton.edu/haushofer/Johannes_Haushofer_CV_of_Failures.pdf [accessed 1 Oct 2017].

33 Erica Falkenström, *Vad vill du – egentligen? Om mål och mening i livet.* [What Do You Want – For Real? On Goals and Meaning in Life] (second revised edition) (Stockholm: Natur och Kultur, 2017), p. 27. Aristotle's arguments can be found in the *Nicomachean Ethics* (Ware, Hertfordshire: Wordsworth Editions, 1996).

INDEX

www.ingramcontent.com/pod-product-compliance
Lightning Source LLC
Chambersburg PA
CBHW020004290326
41935CB00007B/303